Second Edition

Physics in Biology and Medicine

Complementary Science Series

2000/2001

The Physical Basis of Chemistry, 2nd Edition
Warren S. Warren

Physics in Biology and Medicine, 2nd Edition
Paul Davidovits

Introduction to Relativity
John Kogut

Earth Magnetism: A Guided Tour through Magnetic Fields
Wallace Campbell

The Human Genome, 2nd Edition
R. Scott Hawley ► ***Julia Richards*** ► ***Catherine Mori***

1999

Chemistry Connections
Kerry Karukstis ► ***Gerald Van Hecke***

Mathematics for Physical Chemistry, 2nd Edition
Robert Mortimer

Fundamentals of Quantum Mechanics
J.E. House

www.harcourt-ap.com

Second Edition

Physics in Biology and Medicine

Paul Davidovits

Department of Chemistry
Boston College
Chestnut Hill, Massachusetts

An Imprint of Elsevier

San Diego San Francisco New York Boston London Sydney Tokyo

Sponsoring Editor	Jeremy Hayhurst
Production Managers	Joanna Dinsmore and Andre Cuello
Editorial Coordinator	Nora Donaghy
Marketing Manager	Marianne Rutter
Cover Design	Greg Smith
Copyeditor	Sara Black
Proofreader	Cheryl Uppling
Preproduction	Supplinc
Composition	Archetype Publishing, Inc.
Printer	Maple-Vail Book Manufacturing

This book is printed on acid-free paper. ∞

Academic Press
An imprint of Elsevier
525 B Street, Suite 1900, San Diego, California 92101-4495, USA
http://www.academicpress.com

Academic Press
84 Theobald's Road, London WC1X 8RR, UK
http://www.academicpress.com

Library of Congress Catalog Card Number: 00-107669

International Standard Book Number: 0-12-204840-7

PRINTED IN THE UNITED STATES OF AMERICA
04 05 06 07 9 8 7 6 5 4 3

Contents

Preface

Until the mid 1800s it was not clear to what extent the laws of physics and chemistry, which were formulated from the observed behavior of inanimate matter, could be applied to living matter. It was certainly evident that on the large scale the laws were applicable. Animals are clearly subject to the same laws of motion as inanimate objects. The question of applicability arose on a more basic level. Living organisms are very complex. Even a virus, which is one of the simplest biological organisms, consists of millions of interacting atoms. A cell, which is the basic building block of tissue, contains on the average 10^{14} atoms. Living organisms exhibit properties not found in inanimate objects. They grow, reproduce, and decay. These phenomena are so different from the predictable properties of inanimate matter that many scientists in the early 19th century believed that different laws governed the structure and organization of molecules in living matter. Even the physical origin of organic molecules was in question. These molecules tend to be larger and more complex than molecules obtained from inorganic sources. It was thought that the large molecules found in living matter could be produced only by living organisms through a "vital force" that could not be explained by the existing laws of physics. This concept was disproved in 1828 when Friedrich Wöhler synthesized an organic substance, urea, from inorganic chemicals. Soon thereafter many other organic molecules were synthesized without the intervention of biological organisms. Today most scientists believe that there is no special vital force residing in organic substances. Living organisms are governed by the laws of physics on all levels.

Much of the biological research during the past hundred years has been directed toward understanding living systems in terms of basic physical laws. This effort has yielded some significant successes. The atomic structure of many complex biological molecules has now been determined, and the role of these molecules within living systems has been described. It is now possible to explain the functioning of cells and many of their interactions with each other. Yet the work is far from complete. Even when the structure of a complex molecule is known, it is not possible at present to predict its function from its atomic structure. The mechanisms of cell nourishment, growth, reproduction, and communication are still understood only qualitatively. Many of the basic questions in biology remain unanswered. However, biological research has so far not revealed any areas where physical laws do not apply. The amazing properties of life seem to be achieved by the enormously complex organization in living systems.

The aim of this book is to relate some of the concepts in physics to living systems. In general, the text follows topics found in basic college physics texts. The discussion is organized into the following areas: solid mechanics, fluid mechanics, thermodynamics, sound, electricity, optics, and atomic and nuclear physics.

Each chapter contains a brief review of the background physics, but most of the text is devoted to the applications of physics to biology and medicine. No previous knowledge of biology is assumed. The biological systems to be discussed are described in as much detail as is necessary for the physical analysis. Whenever possible, the analysis is quantitative, requiring only basic algebra and trigonometry.

Many biological systems can be analyzed quantitatively. A few examples will illustrate the approach. Under the topic of mechanics we calculate the forces exerted by muscles. We examine the maximum impact a body can sustain without injury. We calculate the height to which a person can jump, and we discuss the effect of an animal's size on the speed at which it can run. In our study of fluids we examine quantitatively the circulation of blood in the body. The theory of fluids allows us also to calculate the role of diffusion in the functioning of cells and the effect of surface tension on the growth of plants in soil. Using the principles of electricity, we analyze quantitatively the conduction of impulses along the nervous system. Each section contains problems that explore and expand some of the concepts.

There are, of course, severe limits on the quantitative application of physics to biological systems. These limitations are discussed.

Many of the advances in the life sciences have been greatly aided by the application of the techniques of physics and engineering to the study of living systems. Some of these techniques are examined in the appropriate sections of the book.

This new edition has been updated and includes a discussion of information theory and descriptions of CT scan and MRI imaging, two techniques that were not available at the writing of the first edition.

A word about units. Most physics and chemistry textbooks now use the MKS International System of units (SI). In practice, however, a variety of units continue to be in use. For example, in the SI system, pressure is expressed in units of pascals (kg/m^2). Both in common use and in the scientific literature one often finds pressure also expressed in units of $dynes/cm^2$, Torr (mm Hg), psi, and atm. In this book I have used mostly SI units. However, other units have also been used when common usage so dictated. In those cases conversion factors have been provided either within the text or in a compilation at the end of Appendix A.

In the first edition of this book I expressed my thanks to W. Chameides, M. D. Egger, L. K. Stark, and J. Taplitz for their help and encouragement. In the writing of this second edition I want to thank Professors R. K. Hobbie and David Cinabro for their careful reading of the manuscript and helpful suggestions. I also appreciate the encouragement and competent direction of J. Hayhurst, S. Stevens, N. Donaghy, and J. Dinsmore at Harcourt/Academic Press.

Paul Davidovits
Chestnut Hill, Massachusetts

Chapter 1

Static Forces

Mechanics is the branch of physics concerned with the effect of forces on the motion of bodies. It was the first branch of physics that was applied successfully to living systems, primarily to understanding the principles governing the movement of animals. Our present concepts of mechanics were formulated by Isaac Newton, whose major work on mechanics, *Principia Mathematica*, was published in 1687. The study of mechanics, however, began much earlier. It can be traced to the Greek philosophers of the fourth century B.C. The early Greeks, who were interested in both science and athletics, were also the first to apply physical principles to animal movements. Aristotle wrote, "The animal that moves makes its change of position by pressing against that which is beneath it. . . . Runners run faster if they swing their arms for in extension of the arms there is a kind of leaning upon the hands and the wrist." Although some of the concepts proposed by the Greek philosophers were wrong, their search for general principles in nature marked the beginning of scientific thought.

After the decline of ancient Greece, the pursuit of all scientific work entered a period of lull that lasted until the Renaissance brought about a resurgence in many activities including science. During this period of revival, Leonardo da Vinci (1452–1519) made detailed observations of animal motions and muscle functions. Since da Vinci, hundreds of people have contributed to our understanding of animal motion in terms of mechanical principles. Their studies have been aided by improved analytic techniques and the development of instruments such as the photographic camera and electronic timers. Today the study of human motion is part of the disciplines of kinesiology, which studies human motion primarily as applied to athletic activities, and biomechanics, a broader area that is concerned not only with muscle movement but also with the physical behavior of bones and organs such as the lungs and the heart. The development of prosthetic devices such as artificial limbs and mechanical hearts is an active area of biomechanical research.

Mechanics, like every other subject in science, starts with a certain number of basic concepts and then supplies the rules by which they are interrelated. Appendix A summarizes the basic concepts in mechanics, providing a review rather than a thorough treatment of the subject. We will now begin our discussion of mechanics by examining static forces that act on the human body. We will first discuss stability and equilibrium of the human body, and then we will calculate the forces exerted by the skeletal muscles on various parts of the body.

1.1 Equilibrium and Stability

The Earth exerts an attractive force on the mass of an object; in fact, every small element of mass in the object is attracted by the Earth. The sum of these forces is the total weight of the body. This weight can be considered a force acting through a single point called the center of mass or center of gravity. As pointed out in Appendix A, a body is in static equilibrium if the vectorial sum of both the forces and the torques acting on the body is zero. If a body is unsupported, the force of gravity accelerates it, and the body is not in equilibrium. In order that a body be in stable equilibrium, it must be properly supported.

The position of the center of mass with respect to the base of support determines whether the body is stable or not. A body is in stable equilibrium under the action of gravity if its center of mass is directly over its base of support (Fig. 1.1). Under this condition, the reaction force at the base of support cancels the force of gravity and the torque produced by it. If the center of mass is outside the base, the torque produced by the weight tends to topple the body (Fig. 1.1c).

The wider the base on which the body rests, the more stable it is; that is, the more difficult it is to topple it. If the wide-based body in Fig. 1.1a is displaced as shown in Fig. 1.2a, the torque produced by its weight tends to restore it to its original position (F_r shown is the reaction force exerted by the surface on the body). The same amount of angular displacement of a narrow-based body results in a torque that will topple it (Fig. 1.2b). Similar considerations show that a body is more stable if its center of gravity is closer to its base.

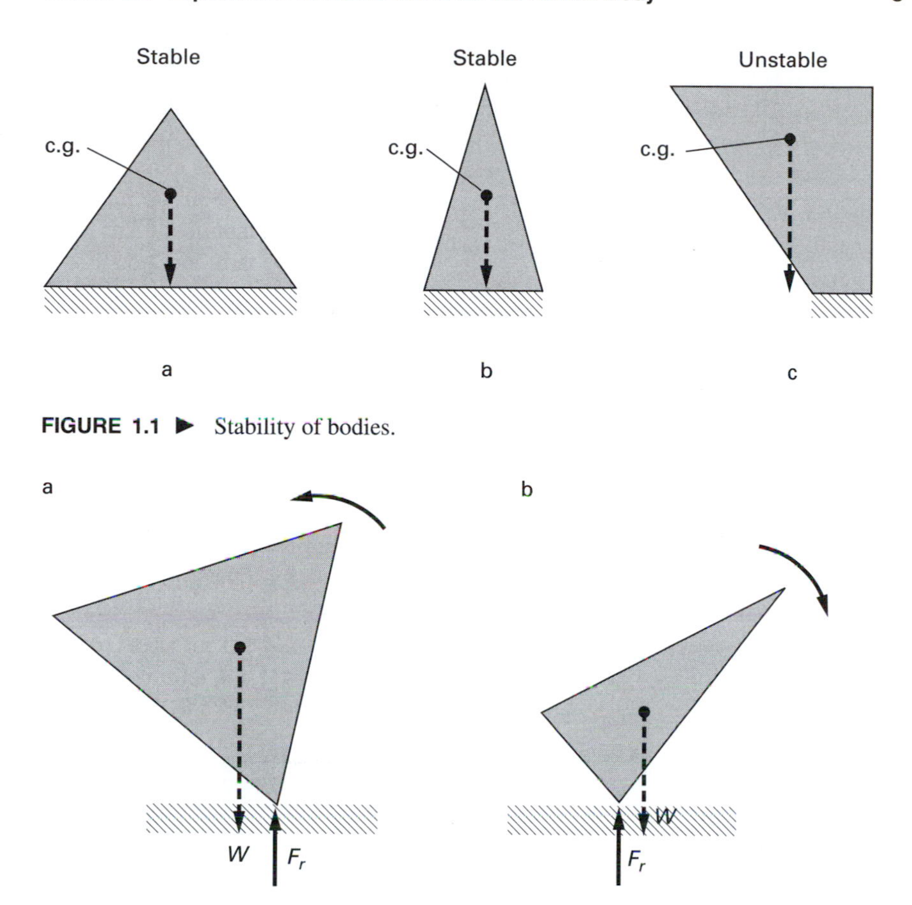

FIGURE 1.1 ▶ Stability of bodies.

FIGURE 1.2 ▶ (a) Torque produced by the weight will restore the body to its original position. (b) Torque produced by the weight will topple the body.

1.2 Equilibrium Considerations for the Human Body

The center of gravity (c.g.) of an erect person with arms at the side is at approximately 56% of the person's height measured from the soles of the feet (Fig. 1.3). The center of gravity shifts as the person moves and bends. The act of balancing requires maintenance of the center of gravity above the feet. A person falls when his center of gravity is displaced beyond the position of the feet.

When carrying an uneven load, the body tends to compensate by bending and extending the limbs so as to shift the center of gravity back over the feet. For example, when a person carries a weight in one arm, the other arm swings away from the body and the torso bends away from the load (Fig. 1.4). This tendency of the body to compensate for uneven weight distribution often

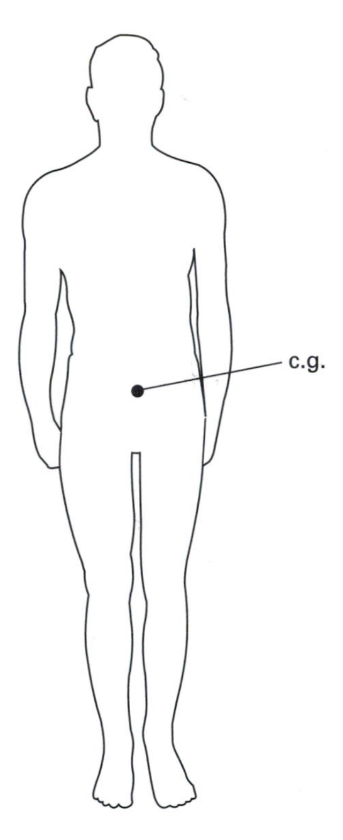

FIGURE 1.3 ▶ Center of gravity for a person.

causes problems for people who have lost an arm, as the continuous compensatory bending of the torso can result in a permanent distortion of the spine. It is often recommended that amputees wear an artificial arm, even if they cannot use it, to restore balanced weight distribution.

1.3 Stability of the Human Body under the Action of an External Force

The body may of course be subject to forces other than the downward force of weight. Let us calculate the magnitude of the force applied to the shoulder that will topple a person standing at rigid attention. The assumed dimensions of the person are as shown in Fig. 1.5. In the absence of the force, the person is in stable equilibrium because his center of mass is above his feet, which are the base of support. The applied force F_a tends to topple the body. When the person topples, he will do so by pivoting around point A—assuming that he does not slide. The counterclockwise torque T_a about this point produced by

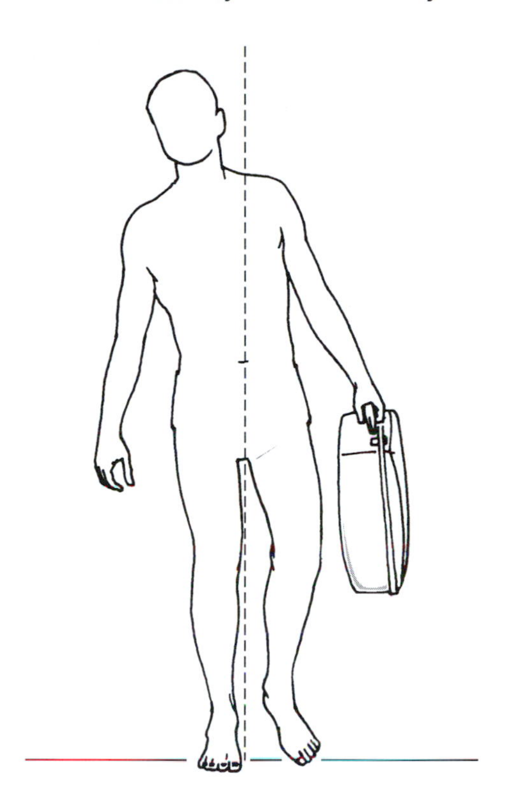

FIGURE 1.4 ► A person carrying a weight.

the applied force is

$$T_a = F_a \times 1.5\ \text{m} \tag{1.1}$$

The opposite restoring torque T_w due to the person's weight is

$$T_w = W \times 0.1\ \text{m} \tag{1.2}$$

Assuming that the mass m of the person is 70 kg, his weight W is

$$W = mg = 70 \times 9.8 = 686\ \text{newton (N)} \tag{1.3}$$

(Here g is the gravitational acceleration, which has the magnitude 9.8 m/sec^2.) The restoring torque produced by the weight is therefore 68.6 newton-meter

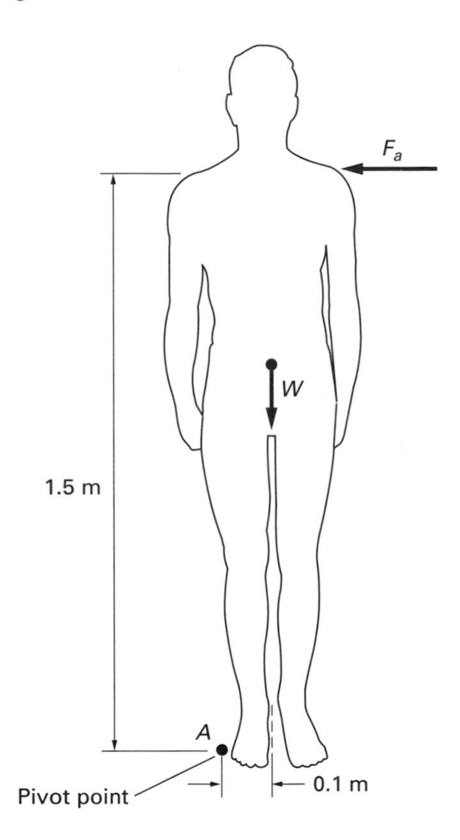

FIGURE 1.5 ▶ A force applied to an erect person.

(N-m). The person is on the verge of toppling when the magnitudes of these two torques are just equal; that is, $T_a = T_w$ or

$$F_a \times 1.5\ \text{m} = 68.6\ \text{N-m} \tag{1.4}$$

Therefore, the force required to topple an erect person is

$$F_a = \frac{68.6}{1.5} = 45.7\ \text{N}\ (10.3\ \text{lb}) \tag{1.5}$$

Actually, a person can withstand a much greater sideways force without losing balance by bending the torso in the direction opposite to the applied force (Fig. 1.6). This shifts the center of gravity away from the pivot point A, increasing the restoring torque produced by the weight of the body.

Stability against a toppling force is also increased by spreading the legs, as shown in Fig. 1.7 and discussed in Exercise 1-1.

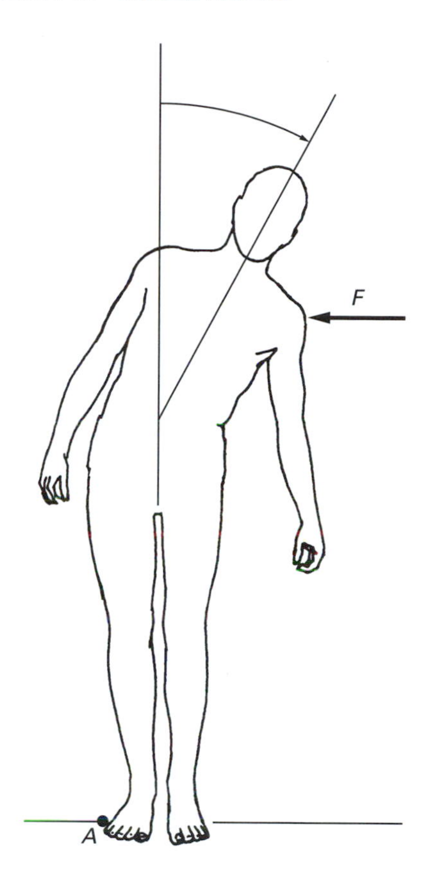

FIGURE 1.6 ▶ Compensating for a side-pushing force.

1.4 Skeletal Muscles

The skeletal muscles producing skeletal movements consist of many thousands of parallel fibers wrapped in a flexible sheath that narrows at both ends into tendons (Fig. 1.8). The tendons, which are made of strong tissue, grow into the bone and attach the muscle to the bone. Most muscles taper to a single tendon. But some muscles end in two or three tendons; these muscles are called, respectively, *biceps* and *triceps*. Each end of the muscle is attached to a different bone. In general, the two bones attached by muscles are free to move with respect to each other at the joints where they contact each other.

This arrangement of muscle and bone was noted by Leonardo da Vinci, who wrote, "The muscles always begin and end in the bones that touch one another, and they never begin and end on the same bone. . . ." He also stated,

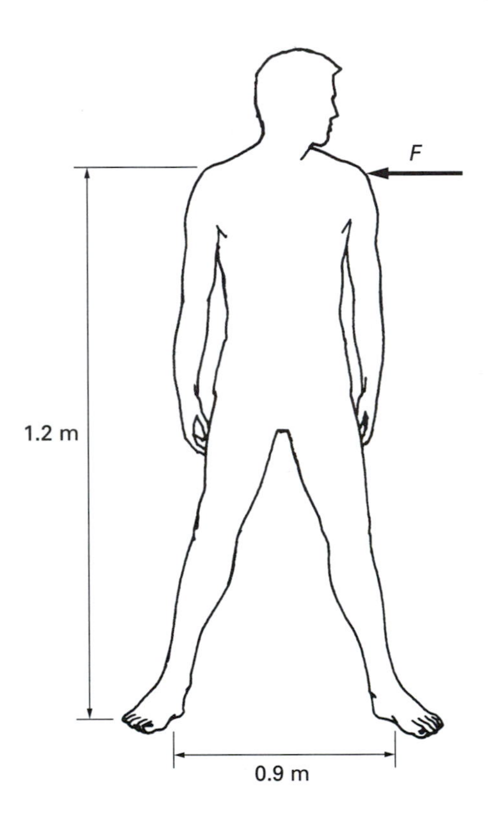

FIGURE 1.7 ▶ Increased stability resulting from spreading the legs.

"It is the function of the muscles to pull and not to push except in the cases of the genital member and the tongue."

Da Vinci's observation about the pulling by muscles is correct. When fibers in the muscle receive an electrical stimulus from the nerve endings that are attached to them, they contract. This results in a shortening of the muscle and a corresponding pulling force on the two bones to which the muscle is attached.

There is a great variability in the pulling force that a given muscle can apply. The force of contraction at any time is determined by the number of individual fibers that are contracting within the muscle. When an individual fiber receives an electrical stimulus, it tends to contract to its full ability. If a stronger pulling force is required, a larger number of fibers are stimulated to contract.

Experiments have shown that the maximum force a muscle is capable of exerting is proportional to its cross section. From measurements, it has been

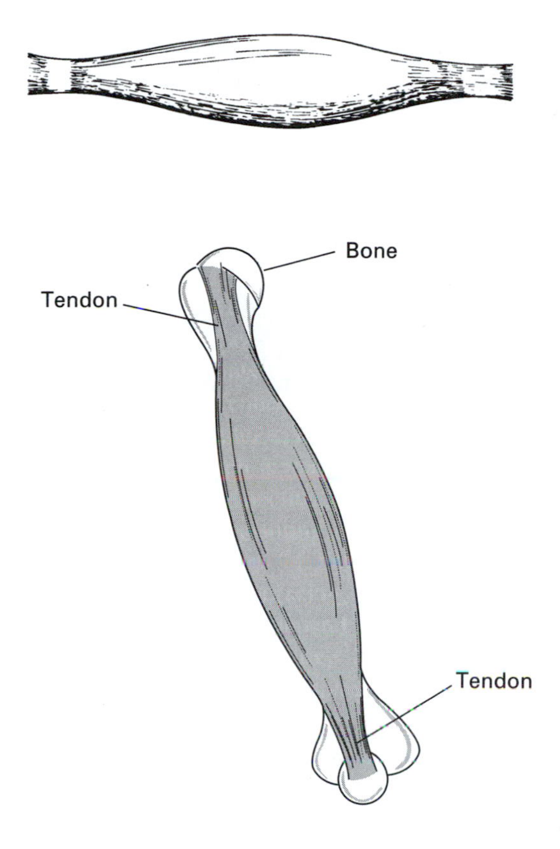

FIGURE 1.8 ▶ Drawing of a muscle.

estimated that a muscle can exert a force of about 7×10^6 dyn/cm^2 of its area (7×10^6 dyn/cm$^2 = 7 \times 10^5$ Pa $= 102$ lb/in^2).

To compute the forces exerted by muscles, the various joints in the body can be conveniently analyzed in terms of levers. Such a representation implies some simplifying assumptions. We will assume that the tendons are connected to the bones at well-defined points and that the joints are frictionless.

Simplifications are often necessary to calculate the behavior of systems in the real world. Seldom are all the properties of the system known, and even when they are known, consideration of all the details is usually not necessary. Calculations are most often based on a model, which is assumed to be a good representation of the real situation.

1.5 Levers

A lever is a rigid bar free to rotate about a fixed point called the *fulcrum*. The position of the fulcrum is fixed so that it is not free to move with respect to the bar. Levers are used to lift loads in an advantageous way and to transfer movement from one point to another.

There are three classes of levers, as shown in Fig. 1.9. In a Class 1 lever, the fulcrum is located between the applied force and the load. A crowbar is an example of a Class 1 lever. In a Class 2 lever, the fulcrum is at one end of the bar; the force is applied to the other end; and the load is situated in between. A wheelbarrow is an example of a Class 2 lever. A Class 3 lever has the fulcrum at one end and the load at the other. The force is applied between the two ends. As we will see, many of the limb movements of animals are performed by Class 3 levers.

It can be shown from the conditions for equilibrium (see Appendix A) that, for all three types of levers, the force F required to balance a load of weight W is given by

$$F = \frac{Wd_1}{d_2}, \tag{1.6}$$

where d_1 and d_2 are the lengths of the lever arms, as shown in Fig. 1.9 (see Exercise 1-2). If d_1 is less than d_2, the force required to balance a load is smaller than the load. The mechanical advantage M of the lever is defined as

$$M = \frac{W}{F} = \frac{d_2}{d_1}. \tag{1.7}$$

Depending on the distances from the fulcrum, the mechanical advantage of a Class 1 lever can be greater or smaller than one. By placing the load close to the fulcrum, with d_1 much smaller than d_2, a very large mechanical advantage can be obtained with a Class 1 lever. In a Class 2 lever, d_1 is always smaller than d_2; therefore, the mechanical advantage of a Class 2 lever is greater than

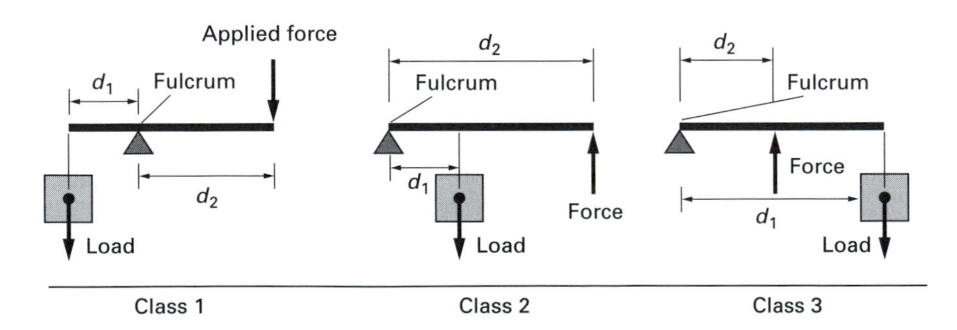

FIGURE 1.9 ▶ The three classes of lever.

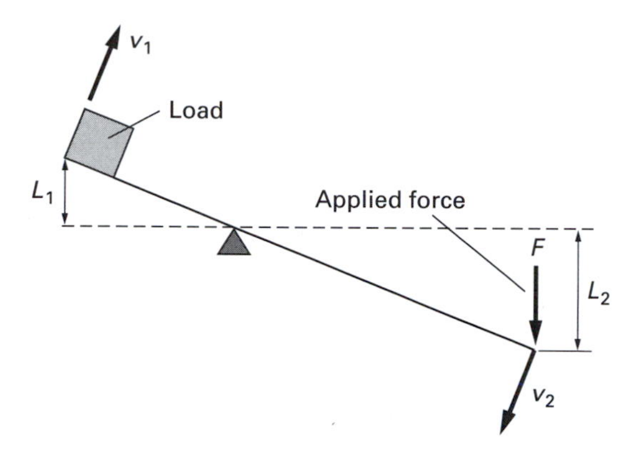

FIGURE 1.10 ▶ Motion of the lever arms in a Class 1 lever.

one. The situation is opposite in a Class 3 lever. Here d_1 is larger than d_2; therefore, the mechanical advantage is always less than one.

A force slightly greater than what is required to balance the load will lift it. As the point at which the force is applied moves through a distance L_2, the load moves a distance L_1 (see Fig. 1.10). The relationship between L_1 and L_2, (see Exercise 1-2) is given by

$$\frac{L_1}{L_2} = \frac{d_1}{d_2}. \tag{1.8}$$

The ratio of velocities of these two points on a moving lever is likewise given by

$$\frac{v_1}{v_2} = \frac{d_1}{d_2}. \tag{1.9}$$

Here v_2 is the velocity of the point where the force is applied, and v_1 is the velocity of the load. These relationships apply to all three classes of levers. Thus, it is evident that the excursion and velocity of the load are inversely proportional to the mechanical advantage.

1.6 The Elbow

The two most important muscles producing elbow movement are the biceps and the triceps (Fig. 1.11). The contraction of the triceps causes an extension, or opening, of the elbow, while contraction of the biceps closes the elbow. In our analysis of the elbow, we will consider the action of only these two muscles. This is a simplification, as many other muscles also play a role in elbow movement. Some of them stabilize the joints at the shoulder as the elbow moves, and others stabilize the elbow itself.

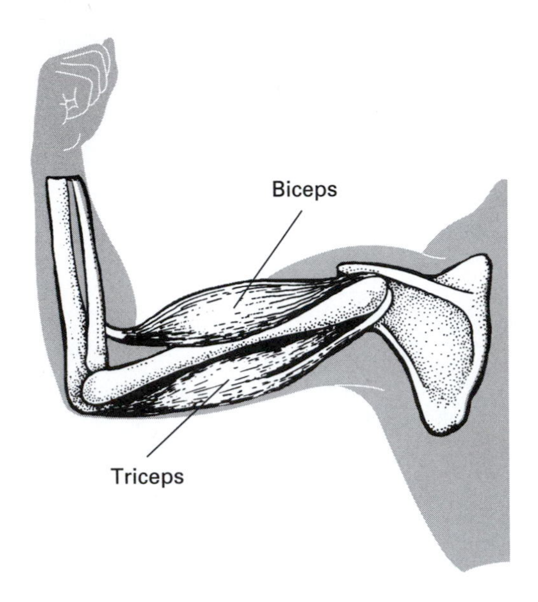

FIGURE 1.11 ▶ The elbow.

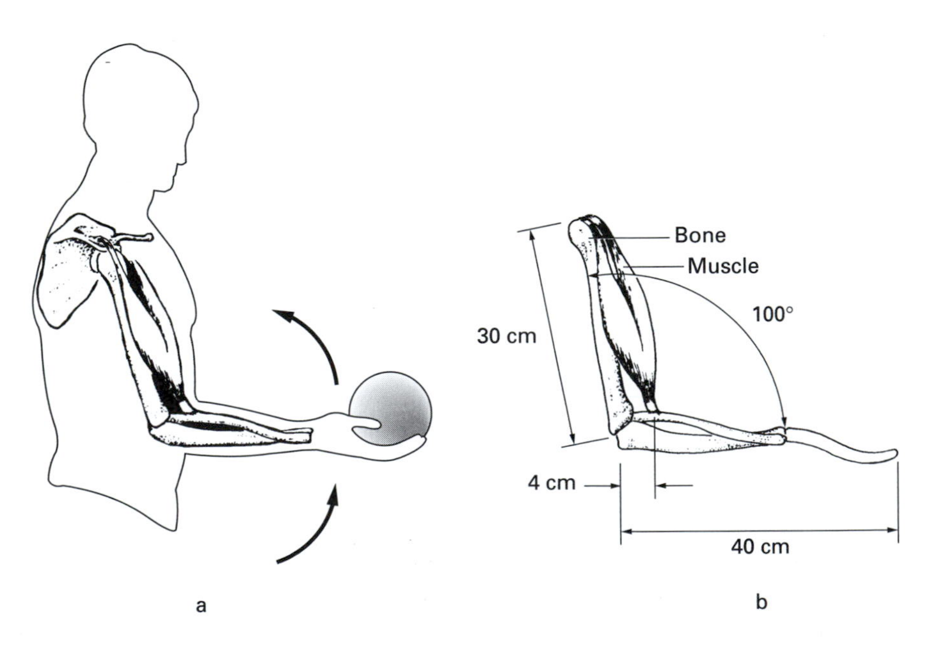

FIGURE 1.12 ▶ (a) Weight held in hand. (b) A simplified drawing of (a).

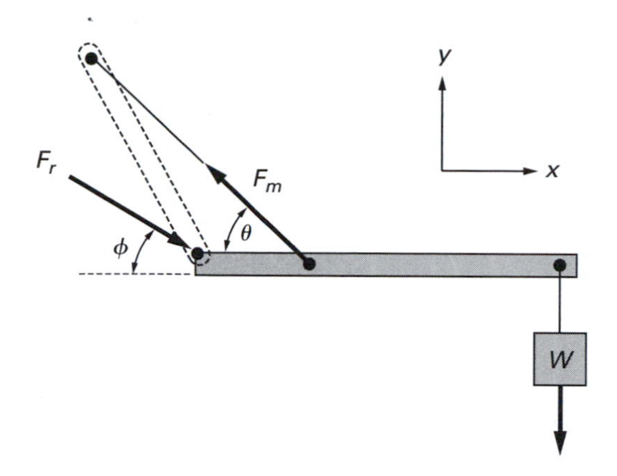

FIGURE 1.13 ▶ Lever representation of Fig. 1.12.

Figure 1.12a shows a weight W held in the hand with the elbow bent at a 100° angle. A simplified diagram of this arm position is shown in Fig. 1.12b. The dimensions shown in Fig. 1.12 are reasonable for a human arm, but they will, of course, vary from person to person. The weight pulls the arm downward. Therefore, the muscle force acting on the lower arm must be in the up direction. Accordingly, the prime active muscle is the biceps. The position of the upper arm is fixed at the shoulder by the action of the shoulder muscles. We will calculate, under the conditions of equilibrium, the pulling force F_m exerted by the biceps muscle and the direction and magnitude of the reaction force F_r at the fulcrum (the joint). The calculations will be performed by considering the arm position as a Class 3 lever, as shown in Fig. 1.13. The x- and y-axes are as shown in Fig. 1.13. The direction of the reaction force F_r shown is a guess. The exact answer will be provided by the calculations.

In this problem we have three unknown quantities: the muscle force F_m, the reaction force at the fulcrum F_r, and the angle, or direction, of this force ϕ. The angle θ of the muscle force can be calculated from trigonometric considerations, without recourse to the conditions of equilibrium. As is shown in Exercise 1-3, the angle θ is 72.6°.

For equilibrium, the sum of the x and y components of the forces must each be zero. From these conditions we obtain

$$x \text{ components of the forces:} \quad F_m \cos\theta = F_r \cos\phi \tag{1.10}$$

$$y \text{ components of the forces:} \quad F_m \sin\theta = W + F_r \sin\phi \tag{1.11}$$

These two equations alone are not sufficient to determine the three unknown quantities. The additional necessary equation is obtained from the torque conditions for equilibrium. In equilibrium, the torque about any point in Fig. 1.13 must be zero. For convenience, we will choose the fulcrum as the point for our torque balance.

The torque about the fulcrum must be zero. There are two torques about this point: a clockwise torque due to the weight and a counterclockwise torque due to the vertical y component of the muscle force. Since the reaction force F_r acts at the fulcrum, it does not produce a torque about this point.

Using the dimensions shown in Fig. 1.12, we obtain

$$4 \text{ cm} \times F_m \sin\theta = 40 \text{ cm} \times W$$

or

$$F_m \sin\theta = 10W \tag{1.12}$$

Therefore, with $\theta = 72.6°$, the muscle force F_m is

$$F_m = \frac{10W}{0.954} = 10.5W \tag{1.13}$$

With a 14-kg (31-lb) weight in hand, the force exerted by the muscle is

$$F_m = 10.5 \times 14 \times 9.8 = 1440 \text{ N (325 lb)}$$

If we assume that the diameter of the biceps is 8 cm and that the muscle can produce a 7×10^6 dyn force for each square centimeter of area, the arm is capable of supporting a maximum of 334 N (75 lb) in the position shown in Fig. 1.13 (see Exercise 1-4).

The solutions of Eqs. 1.10 and 1.11 provide the magnitude and direction of the reaction force F_r. Assuming as before that the weight supported is 14 kg, these equations become

$$\begin{aligned} 1440 \times \cos 72.6 &= F_r \cos\phi \\ 1440 \times \sin 72.6 &= 14 \times 9.8 + F_r \sin\phi \end{aligned} \tag{1.14}$$

or

$$\begin{aligned} F_r \cos\phi &= 430 \text{ N} \\ F_r \sin\phi &= 1240 \text{ N} \end{aligned} \tag{1.15}$$

Squaring both equations, using $\cos^2\phi + \sin^2\phi = 1$ and adding them, we obtain

$$F_r^2 = 1.74 \times 10^6 \text{ N}^2$$

or

$$F_r = 1320 \text{ N (298 lb)} \tag{1.16}$$

From Eqs. 1.14 and 1.15, the cotangent of the angle is

$$\cot\phi = \frac{430}{1240} = 0.347 \tag{1.17}$$

and

$$\phi = 70.9°$$

Exercises 1-5, 1-6, and 1-7 present other similar aspects of biceps mechanics. In these calculations we have omitted the weight of the arm itself, but this effect is considered in Exercise 1-8. The forces produced by the triceps muscle are examined in Exercise 1-9.

Our calculations show that the forces exerted on the joint and by the muscle are large. In fact, the force exerted by the muscle is much greater than the weight it holds up. This is the case with all the skeletal muscles in the body. They all apply forces by means of levers that have a mechanical advantage less than one. As mentioned earlier, this arrangement provides for greater speed of the limbs. A small change in the length of the muscle produces a relatively larger displacement of the limb extremities (see Exercise 1-10). It seems that nature prefers speed to strength. In fact, the speeds attainable at limb extremities are remarkable. A skilled pitcher can hurl a baseball at a speed in excess of 100 mph. Of course, this is also the speed of his hand at the point where he releases the ball.

1.7 The Hip

Figure 1.14 shows the hip joint and its simplified lever representation, giving dimensions that are typical for a male body. The hip is stabilized in its socket by a group of muscles, which is represented in Fig. 1.14b as a single resultant force F_m. When a person stands erect, the angle of this force is about 71° with respect to the horizon. W_L represents the combined weight of the leg, foot, and thigh. Typically, this weight is a fraction (0.185) of the total body weight W (i.e., $W_L = 0.185W$). The weight W_L is assumed to act vertically downward at the midpoint of the limb.

We will now calculate the magnitude of the muscle force F_m and the force F_R at the hip joint when the person is standing erect on one foot as in a slow walk, as shown in Fig. 1.14. The force W acting on the bottom of the lever is the reaction force of the ground on the foot of the person. This is the force that supports the weight of the body.

From equilibrium conditions, using the procedure outlined in Section 1.6, we obtain

$$F_m \cos 71^\circ - F_R \cos\theta = 0 \quad (x \text{ components of the force} = 0) \tag{1.18}$$

$$F_m \sin 71^\circ + W - W_L - F_R \sin\theta = 0 \quad (y \text{ components of the force} = 0) \tag{1.19}$$

$$(F_R \sin\theta) \times 7\text{ cm} + W_L \times 10\text{ cm} - W \times 18\text{ cm} = 0 \quad (\text{torque about point } A = 0) \tag{1.20}$$

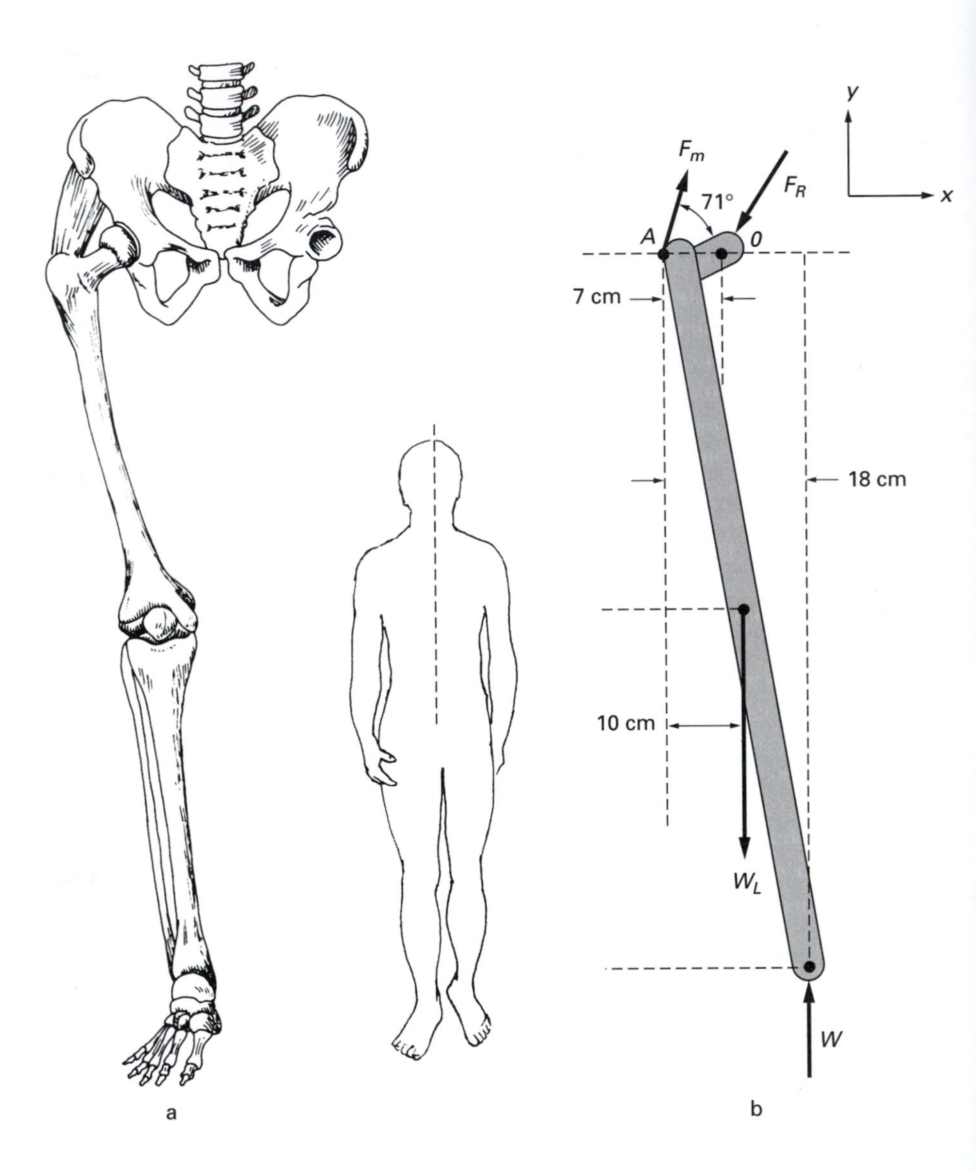

FIGURE 1.14 ▶ (a) The hip. (b) Its lever representation.

Since $W_L = 0.185W$, from Eq. 1.20 we have

$$F_R \sin\theta = 2.31W$$

Using the result in Eq. 1.19, we obtain

$$F_m = \frac{1.50W}{\sin 71^\circ} = 1.59W \tag{1.21}$$

From Eq. 1.18, we obtain

$$F_R \cos\theta = 1.59W \cos 71^\circ = 0.52W$$

therefore,

$$\theta = \tan^{-1} 4.44 = 77.3^\circ$$

and

$$F_R = 2.37W \tag{1.22}$$

This calculation shows that the force on the hip joint is nearly two and one-half times the weight of the person. Consider, for example, a person whose mass is 70 kg and weight is $9.8 \times 70 = 686$ N (154 lb). The force on the hip joint is 1625 N (366 lb).

1.7.1 Limping

Persons who have an injured hip limp by leaning toward the injured side as they step on that foot (Fig. 1.15). As a result, the center of gravity of the body shifts into a position more directly above the hip joint, decreasing the force on the injured area. Calculations for the case in Fig. 1.15 show that the muscle force $F_m = 0.47W$ and that the force on the hip joint is $1.28W$ (see Exercise 1-11). This is a significant reduction from the forces applied during a normal one-legged stance.

1.8 The Back

When the trunk is bent forward, the spine pivots mainly on the fifth lumbar vertebra (Fig. 1.16a). We will analyze the forces involved when the trunk is bent at 60° from the vertical with the arms hanging freely. The lever model representing the situation is given in Fig. 1.16.

The pivot point A is the fifth lumbar vertebra. The lever arm AB represents the back. The weight of the trunk W_1 is uniformly distributed along the back; its effect can be represented by a weight suspended in the middle. The weight of the head and arms is represented by W_2 suspended at the end of the lever arm. The erector spinalis muscle, shown as the connection D-C attached at a point two-thirds up the spine, maintains the position of the back. The angle between the spine and this muscle is about 12°. For a 70-kg man, W_1 and W_2 are typically 320 N (72 lb) and 160 N (36 lb), respectively.

Solution of the problem is left as an exercise. It shows that just to hold up the body weight, the muscle must exert a force of 2000 N (450 lb) and the compressional force of the fifth lumbar vertebra is 2230 N (500 lb). If, in addition, the person holds a 20-kg weight in his hand, the force on the muscle

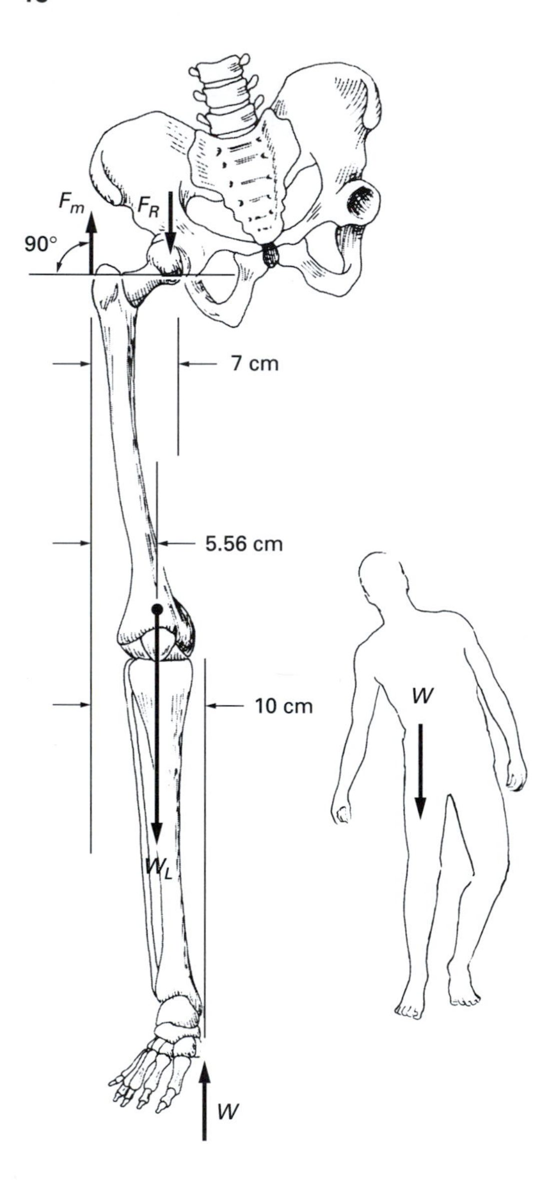

FIGURE 1.15 ▶ Walking on an injured hip.

is 3220 N (725 lb), and the compression of the vertebra is 3490 N (785 lb). (See Exercise 1-12.)

This example indicates that large forces are exerted on the fifth lumbar vertebra. It is not surprising that backaches originate most frequently at this point. It is evident too that the position shown in the figure is not the recommended way of lifting a weight.

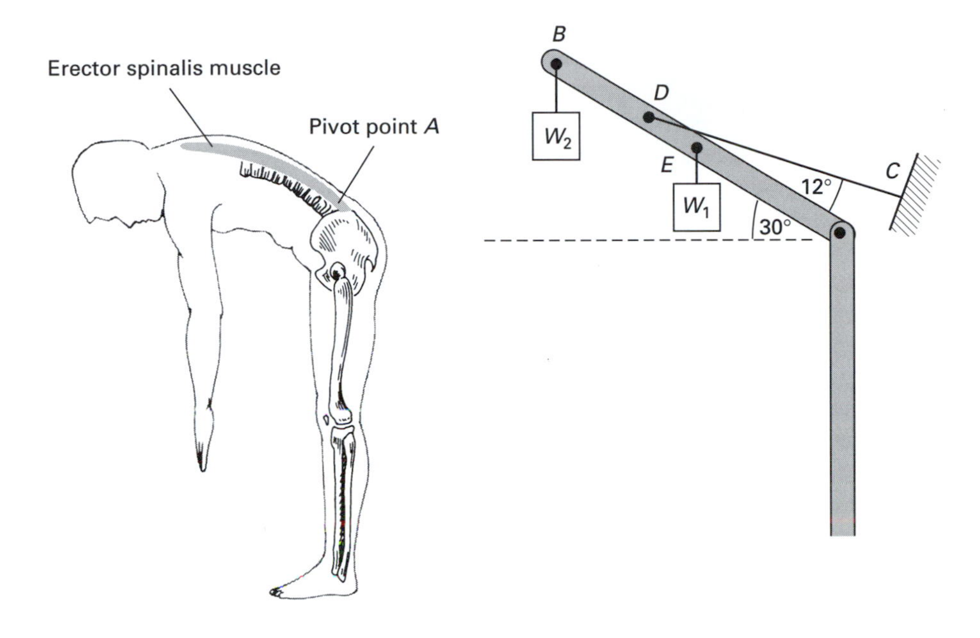

FIGURE 1.16 ▶ (Left) The bent back. (Right) Lever representation.

1.9 Standing Tip-Toe on One Foot

The position of the foot when standing on tiptoe is shown in Fig. 1.17. The total weight of the person is supported by the reaction force at point A. This is a Class 1 lever with the fulcrum at the contact of the tibia. The balancing force is provided by the muscle connected to the heel by the Achilles tendon.

The dimensions and angles shown in Fig. 1.17b are reasonable values for this situation. Calculations show that while standing tiptoe on one foot the compressional force on the tibia is $3.5W$ and the tension force on the Achilles tendon is $2.5 \times W$. (See Exercise 1-13.) Standing on tiptoe is a fairly strenuous position.

1.10 Dynamic Aspects of Posture

In our treatment of the human body, we have assumed that the forces exerted by the skeletal muscles are static. That is, they are constant in time. In fact, the human body (and bodies of all animals) is a dynamic system continually responding to stimuli generated internally and by the external environment. As has been experimentally demonstrated, even an apparently simple act of

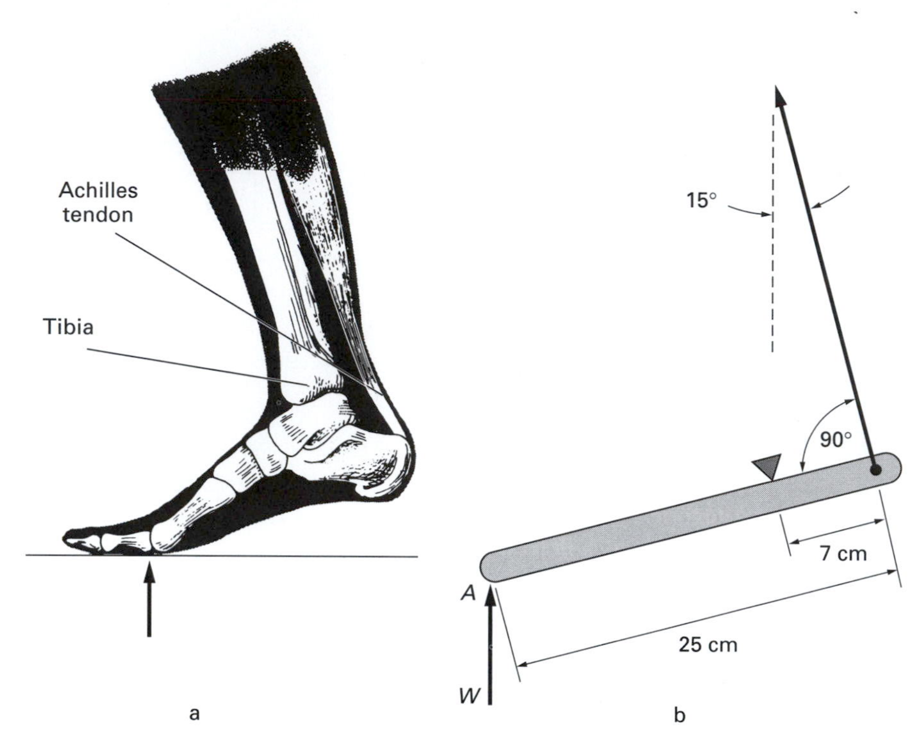

FIGURE 1.17 ▶ (a) Standing on tip-toe. (b) Lever model.

standing upright requires the body to be in a continual back and forth, left right, swaying motion to maintain the posture.

Another aspect of the body dynamics is the interconnectedness of the musculoskeletal system. Through one path or another, all muscles and bones are connected to one another, and a change in muscle tension or limb position in one part of the body must be accompanied by a compensating change elsewhere. The system can be visualized as a complex tentlike structure. The bones act as the tent poles and the muscles as the ropes bringing into and balancing the body in the desired posture. The proper functioning of this type of a structure requires that the forces be appropriately distributed over all the bones and muscles. In a tent, when the forward-pulling ropes are tightened, the tension in the back ropes must be correspondingly increased; otherwise, the tent collapses in the forward direction. The musculoskeletal system operates in an analogous way. For example, excessive tightness, perhaps through overexertion, of the large muscles at the front of our legs will tend to pull the torso forward. To compensate for this forward pull, the muscles in the back must also tighten, often exerting excess force on the more delicate structures of the lower back. In this way, excess tension in one set of muscles may be reflected as pain in an entirely different part of the body.

▶ EXERCISES ▶

1-1. (a) Explain why the stability of a person against a toppling force is increased by spreading the legs as shown in Fig. 1.7. (b) Calculate the force required to topple a person standing with his feet spread 0.9 m apart as shown in Fig. 1.7. Assume the person does not slide.

1-2. Derive the relationships stated in Eqs. 1.6, 1.7, and 1.8.

1-3. Using trigonometry, calculate the angle θ in Fig. 1.13. The dimensions are specified in Fig. 1.12b.

1-4. Using the data provided in the text, calculate the maximum weight that the arm can support in the position shown in Fig. 1.12.

1-5. Calculate the force applied by the biceps and the reaction force at the joint as a result of a 14-kg weight held in hand when the elbow is at (a) 160° and (b) 60°.

Assume that the upper part of the arm remains fixed as in Fig. 1.12. Note that under these conditions the lower part of the arm is no longer horizontal.

1-6. Consider again Fig. 1.12. Now let the 14-kg weight hang from the middle of the lower arm (20 cm from the fulcrum). Calculate the biceps force and the reaction force at the joint.

1-7. Consider the situation when the arm in Fig. 1.13 supports two 14-kg weights, one held by the hand as in Fig. 1.13 and the other supported in the middle of the arm as in Exercise 1-6. (a) Calculate the force of the biceps muscle and the reaction force. (b) Are the forces calculated in part (a) the same as the sum of the forces produced when the weights are suspended individually?

1-8. Calculate the additional forces due to the weight of the arm itself in Fig. 1.13. Assume that the lower part of the arm has a mass of 2 kg and that its total weight can be considered to act at the middle of the lower arm, as in Exercise 1-6.

1-9. Estimate the dimensions of your own arm, and draw a lever model for the extension of the elbow by the triceps. Calculate the force of the triceps in a one arm push-up in a hold position at an elbow angle of 100°.

1-10. Suppose that the biceps in Fig. 1.13 contracts 2 cm. What is the upward displacement of the weight? Suppose that the muscle contraction is uniform in time and occurs in an interval of 0.5 sec. Compute the velocity of the point of attachment of the tendon to the bone and the velocity of the weight. Compare the ratio of the velocities to the mechanical advantage.

1-11. Calculate the forces in the limping situation shown in Fig. 1.15. At what angle does the force F_R act?

1-12. (a) Calculate the force exerted by the muscle and the compression force on the fifth lumbar vertebra in Fig. 1.16. (b) Repeat the calculations in (a) for the case when the person shown in Fig. 1.16 holds a 20-kg weight in his hand.

1-13. Calculate the force on the tibia and on the Achilles tendon in Fig. 1.17.

Chapter 2

Friction

If we examine the surface of any object, we observe that it is irregular. It has protrusions and valleys. Even surfaces that appear smooth to the eye show such irregularities under microscopic examination. When two surfaces are in contact, their irregularities intermesh, and as a result there is a resistance to the sliding or moving of one surface on the other. This resistance is called *friction*. If one surface is to be moved with respect to another, a force has to be applied to overcome friction.

Consider a block resting on a surface as shown in Fig. 2.1. If we apply a force F to the block, it will tend to move. But the intermeshing of surfaces produces a frictional reaction force F_f that opposes motion. In order to move the object along the surface, the applied force must overcome the frictional force. The magnitude of the frictional force depends on the nature of the surfaces; clearly, the rougher the surfaces, the greater is the frictional force. The frictional property of the surfaces is represented by the coefficient of friction μ. The magnitude of the frictional force depends also on the force F_n perpendicular to the surfaces that presses the surfaces together. The magnitude of the force that presses the surfaces together determines to what extent the irregularities are intermeshed.

The frictional force F_f is given by

$$F_f = \mu F_n \tag{2.1}$$

In general, the frictional force on an object in contact with a surface is somewhat larger when the object is stationary than when it is moving. In other words, it takes a larger force to get the object moving against a frictional force

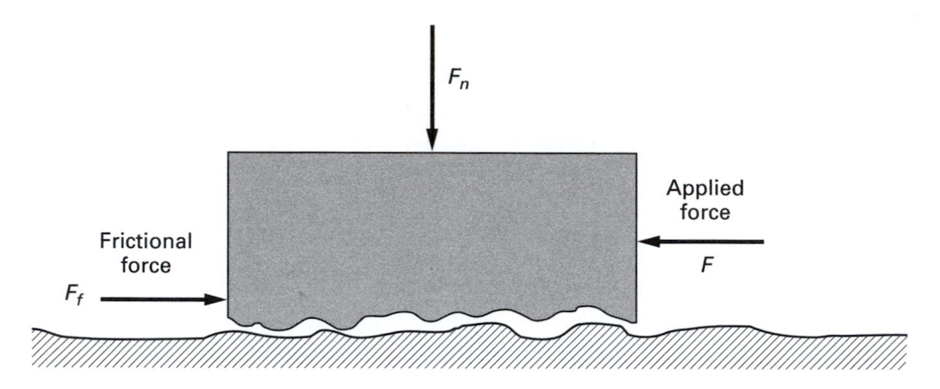

FIGURE 2.1 ▶ Friction.

than to keep it in motion. This is not surprising because in the stationary case the irregularities of the two surfaces can settle more deeply into each other.

The magnitude of the frictional force does not depend on the size of the contact area. If the surface contact area is increased, the force per unit area (pressure) is decreased, and this reduces the interpenetration of the irregularities. However, at the same time, the number of irregularities is proportionately increased. As a result, the total frictional force is unchanged. Coefficients of friction between some surfaces are shown in Table 2.1.

We have illustrated the concept of friction with surfaces sliding along each other, but frictional forces are encountered also in rolling (rolling friction) and in fluid flows (viscous friction). Rolling motion is not encountered in living systems, but viscous friction plays an important role in the flow of blood and other biological fluids.

Whereas sliding friction is independent of velocity, fluid friction has a strong velocity dependence. We will discuss this in Chapter 3.

Friction is everywhere around us. It is both a nuisance and an indispensable factor in the ability of animals to move. Without friction an object that is pushed into motion would continue to move forever (Newton's first law,

TABLE 2.1 ▶ Coefficients of Friction, Static (μ_s) and Kinetic (μ_k)

Surfaces	μ_k	μ_s
Leather on oak	0.6	0.5
Rubber on dry concrete	0.9	0.7
Steel on ice	0.02	0.01
Dry bone on bone		0 3
Bone on joint, lubricated		0.003

Appendix A). The slightest force would send us into eternal motion. It is the frictional force that dissipates kinetic energy into heat and eventually stops the object. (See Exercise 2-1.) Without friction we could not walk; nor could we balance on an inclined plane. (See Exercise 2-2.) In both cases, friction provides the necessary reaction force. Friction also produces undesirable wear and tear and destructive heating of contact surfaces. Both nature and engineers attempt to maximize friction where it is necessary and minimize it where it is destructive. Friction is greatly reduced by introducing a fluid such as oil at the interface of two surfaces. The fluid fills the irregularities and therefore smooths out the surfaces. A natural example of such lubrication occurs in the joints of animals, which are lubricated by a fluid called the synovial fluid. This lubricant reduces the coefficient of friction by about a factor of 100. As is evident from Table 2.1, nature provides very efficient joint lubrication. The coefficient of friction here is significantly lower than for steel on ice.

We will illustrate the effects of friction with a few examples.

2.1 Standing at an Incline

Referring to Fig. 2.2, let us calculate the angle of incline θ of an oak board on which a person of weight W can stand without sliding down. Assume that she is wearing leather-soled shoes and that she is standing in a vertical position as shown in the figure.

The force F_n normal to the inclined surface is

$$F_n = W \cos\theta \tag{2.2}$$

The static frictional force F_f is

$$F_f = \mu F_n = \mu_s W \cos\theta = 0.6W \cos\theta \tag{2.3}$$

The force parallel to the surface F_p, which tends to cause the sliding, is

$$F_p = W \sin\theta \tag{2.4}$$

The person will slide when the force F_p is greater than the frictional force F_f; that is,

$$F_p > F_f \tag{2.5}$$

At the onset of sliding, these two forces are just equal; therefore,

$$\begin{aligned} F_f &= F_p \\ 0.6W \cos\theta &= W \sin\theta \end{aligned} \tag{2.6}$$

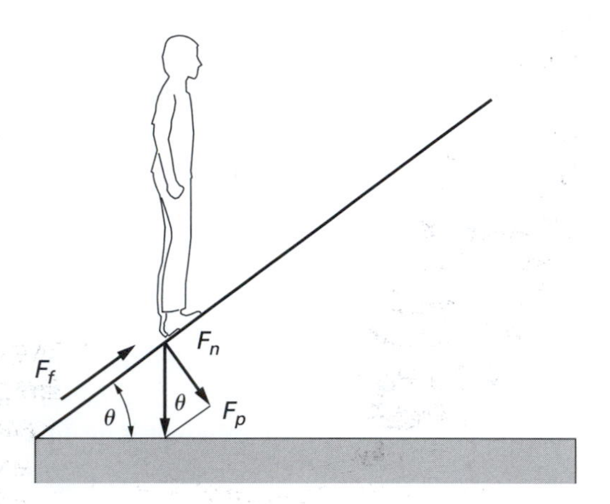

FIGURE 2.2 ▶ Standing on an incline.

or

$$\frac{\sin\theta}{\cos\theta} = \tan\theta = 0.6$$

Therefore $\theta = 31^\circ$.

2.2 Friction at the Hip Joint

We have shown in Chapter 1 that the forces acting on the joints are very large. When the joints are in motion, these large forces produce frictional wear, which could be damaging unless the joints are well lubricated. Frictional wear at the joints is greatly reduced by a smooth cartilage coating at the contact ends of the bone and by synovial fluid which lubricates the contact areas. We will now examine the effect of lubrication on the hip joint in a person. When a person walks, the full weight of the body rests on one leg through most of each step. Because the center of gravity is not directly above the joint, the force on the joint is greater than the weight. Depending on the speed of walking, this force is about 2.4 times the weight (see Chapter 1). In each step, the joint rotates through about 60°. Since the radius of the joint is about 3 cm, the joint slides about 3 cm inside the socket during each step. The frictional force on the joint is

$$F_f = 2.4W\mu \tag{2.7}$$

The work expended in sliding the joint against this friction is the product of the frictional force and the distance over which the force acts (see Appendix A). Thus, the work expended during each step is

$$\text{Work} = F_f \times \text{distance} = 2.4W\mu(3\text{ cm}) = 7.2\mu W \text{ erg} \tag{2.8}$$

If the joint were not lubricated, the coefficient of friction (μ) would be about 0.3. Under these conditions, the work expended would be

$$\text{Work} = 2.16 \times W \text{ erg} \tag{2.9}$$

This is a large amount of work to expend on each step. It is equivalent to lifting the full weight of the person 2.16 cm. Furthermore, this work would be dissipated into heat energy, which would destroy the joint.

As it is, the joint is well lubricated, and the coefficient of friction is only 0.003. Therefore, the work expended in counteracting friction and the resultant heating of the joint are negligible. However, as we age, the joint cartilage begins to wear, efficiency of lubrication decreases, and the joints may become seriously damaged. Studies indicate that by the age of 70 about two-thirds of people have knee joint problems and about one-third have hip problems.

2.3 Spine Fin of a Catfish

Although in most cases good lubrication of bone-contact surfaces is essential, there are a few cases in nature where bone contacts are purposely unlubricated to increase friction. The catfish has such a joint connecting its dorsal spine fin to the rest of its skeleton (Fig. 2.3). Normally the fin is folded flat against the body, but when the fish is attacked, the appropriate muscles pull the bone of the fin into a space provided in the underlying skeleton. Since the coefficient of friction between the fin bone and the skeleton is high, the frictional force tends to lock the fin in the up position. In order to remove the fin, a force must be applied in a predominantly vertical direction with respect to the underlying skeleton. The erect sharp fin discourages predators from eating the catfish.

Figure 2.3b is a simplified representation of the spine and the protruding fin. The shaded block represents the movable fin bone, and the horizontal block is the spine holding the fin. Assume that a force F at an angle θ is applied at point A to dislodge the bone. The force is shown to act at point A, 2.5 cm above point B. The dimensions shown in the figure are to be used in the calculations required for Exercise 2-3. The applied force tips the bone, and as a result reaction forces are set up at points B and C. The components of these forces normal to the fin-bone surface produce frictional forces that resist removal of the bone. Calculation of some of the properties of the locking mechanism is left as an exercise.

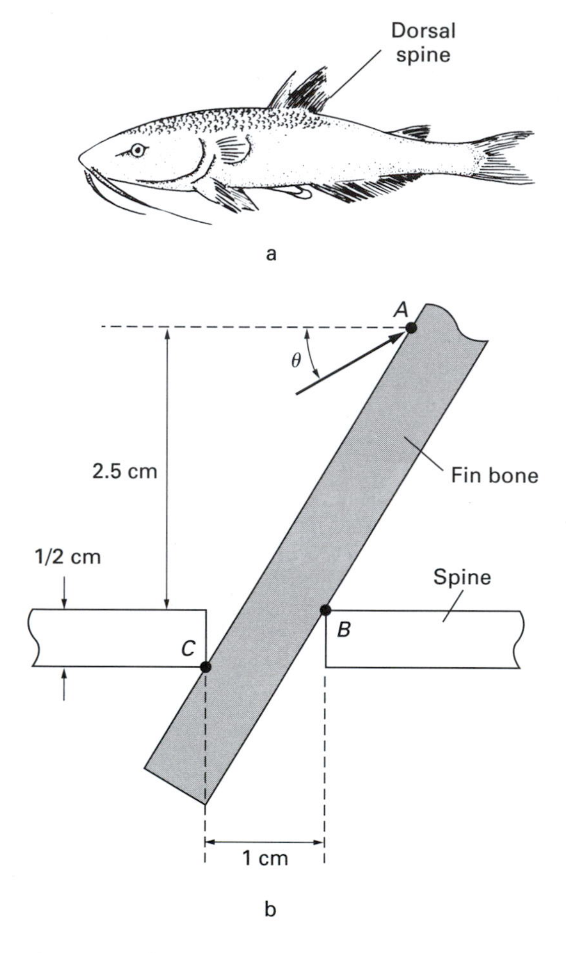

FIGURE 2.3 ▶ (a) Catfish. (b) Simplified representation of the spine in the catfish.

▶ EXERCISES ▶

2-1. (a) Assume that a 50-kg skater, on level ice, has built up her speed to 30 km/h. How far will she coast before the sliding friction dissipates her energy? (Kinetic energy $= \frac{1}{2}mv^2$; see Appendix A.) (b) How does the distance of coasting depend on the mass of the skater?

2-2. Referring to Fig. 1.5, compute the coefficient of friction at which the tendency of the body to slide and the tendency to topple due to the applied force are equal.

2-3. (a) Referring to Fig. 2.3, assume that a dislodging force of 0.1 N is applied at $\theta = 45°$ and the angle between the fin bone and the spine is 20°. Calculate the minimum value for the coefficient of friction between the bones to prevent dislodging of the bone. (b) Assuming that the coefficient of friction is 0.3, what is the value of the angle θ at which a force of 0.2 N will just dislodge the bone? What would this angle be if the bones were lubricated ($\mu = 0.003$)?

 Chapter 3

Translational Motion

In general, the motion of a body can be described in terms of *translational* and *rotational* motion. In pure translational motion all parts of the body have the same velocity and acceleration (Fig. 3.1). In pure rotational motion, such as the rotation of a bar around a pivot, the rate of change in the angle θ is the same for all parts of the body (Fig. 3.2), but the velocity and acceleration along the body depend on the distance from the center of rotation. Many motions and movements encountered in nature are combinations of rotation and translation, as in the case of a body that rotates while falling. It is convenient, however, to discuss these motions separately. In this chapter, we discuss translation. Rotation is discussed in the following chapter.

The equations of translational motion for constant acceleration are presented in Appendix A and may be summarized as follows: In uniform acceleration, the final velocity (v) of an object that has been accelerated for a time t is

$$v = v_0 + at \tag{3.1}$$

Here v_0 is the initial velocity of the object and a is the acceleration. Acceleration can therefore be expressed as

$$a = \frac{v - v_0}{t} \tag{3.2}$$

The average velocity during the time interval t is

$$v_{av} = \frac{v + v_0}{2} \tag{3.3}$$

The distance s traversed during this time is

$$s = v_{av}t \tag{3.4}$$

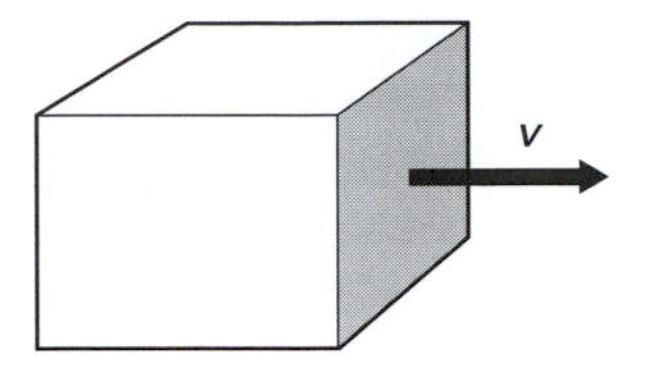

FIGURE 3.1 ► Translational motion.

Using Eqs. 3.1 and 3.2, we obtain

$$s = v_0 t = \frac{at^2}{2} \tag{3.5}$$

By substituting $t = (v - v_0)/a$ from Eq. 3.1 into Eq. 3.5, we obtain

$$v^2 = v_0^2 + 2as \tag{3.6}$$

Let us now apply these equations to some problems in the life sciences. Most of our calculations will relate to various aspects of jumping. Although in the process of jumping the acceleration of the body is usually not constant, the assumption of constant acceleration is necessary to solve the problems without undue difficulties.

3.1 Vertical Jump

Consider a simple vertical jump in which the jumper starts in a crouched position and then pushes off with her feet (Fig. 3.3).

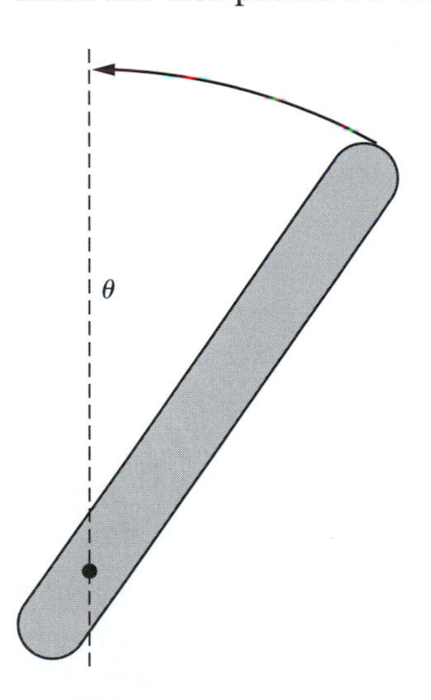

FIGURE 3.2 ► Rotational motion.

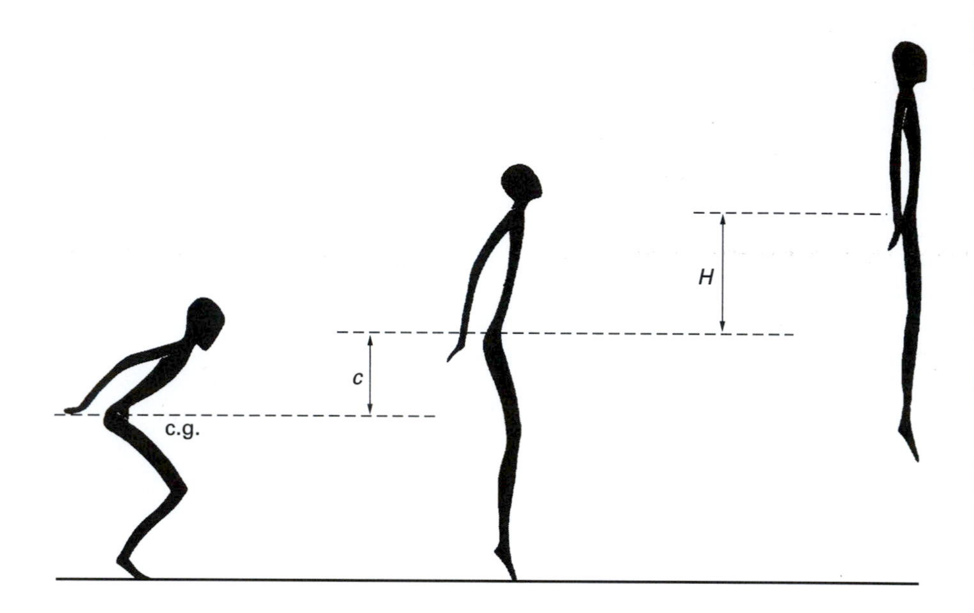

FIGURE 3.3 ▶ Vertical jump.

We will calculate here the height H attained by the jumper. In the crouched position, at the start of the jump, the center of gravity is lowered by a distance c. During the act of jumping, the legs generate a force by pressing down on the surface. Although this force varies through the jump, we will assume that it has a constant average value F.

Because the feet of the jumper exert a force on the surface, an equal upward-directed force is exerted by the surface on the jumper (Newton's third law). Thus, there are two forces acting on the jumper: her weight (W), which is in the downward direction, and the reaction force (F), which is in the upward direction. The net upward force on the jumper is $F - W$ (see Fig. 3.4). This force acts on the jumper until her body is erect and her feet leave the ground. The upward force, therefore, acts on the jumper through a distance c (see Fig. 3.3). The acceleration of the jumper in this stage of the jump (see Appendix A) is

$$a = \frac{F - W}{m} = \frac{F - W}{W/g} \tag{3.7}$$

where W is the weight of the jumper and g is the gravitational acceleration. A consideration of the forces acting on the Earth (Fig. 3.5) shows that an equal force accelerates the Earth in the opposite direction. However, the mass of the Earth is so large that its acceleration due to the jump is negligible.

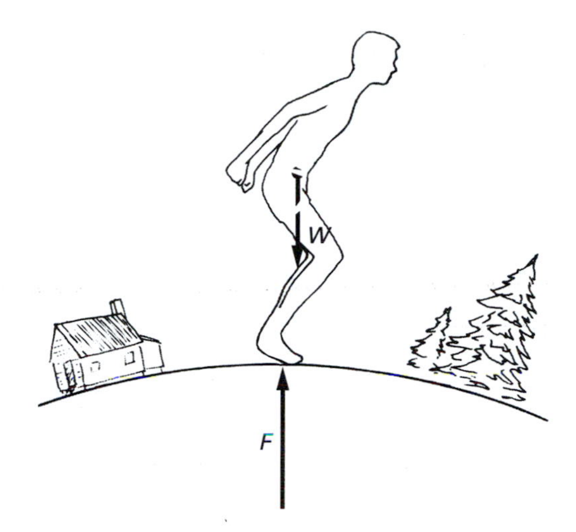

FIGURE 3.4 ▶ Forces on the jumper.

The acceleration shown in Eq. 3.7 takes place over a distance c. Therefore, the velocity v of the jumper at take-off as given by Eq. 3.6 is

$$v^2 = v_0^2 + 2ac \tag{3.8}$$

Since the initial velocity at the start of the jump is zero (i.e., $v_0 = 0$), the take-off velocity is

$$v^2 = \frac{2(F - W)c}{W/g} \tag{3.9}$$

(Here we have substituted

$$a = \frac{F - W}{W/g}$$

into Eq. 3.8.)

After the body leaves the ground, the only force acting on it is the force of gravity W, which produces a downward acceleration $-g$ on the body. At the

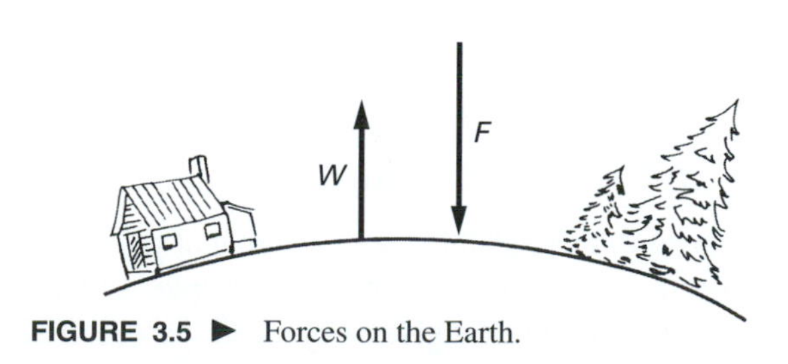

FIGURE 3.5 ▶ Forces on the Earth.

maximum height H, just before the body starts falling back to the ground, the velocity is zero. The initial velocity for this part of the jump is the take-off velocity v given by Eq. 3.9. Therefore, from Eq. 3.6, we obtain

$$0 = \frac{2(F - W)c}{W/g} - 2gH \tag{3.10}$$

From this, the height of the jump is

$$H = \frac{(F - W)c}{W} \tag{3.11}$$

Now let us estimate the numerical value for the height of the jump. Experiments have shown that in a good jump a well-built person generates an average reaction force that is twice his/her weight (i.e., $F = 2W$). In that case, the height of jump is $H = c$. The distance c, which is the lowering of the center of gravity in the crouch, is proportional to the length of the legs. For an average person, this distance is about 60 cm, which is our estimate for the height of a vertical jump.

The height of a vertical jump can also be computed very simply from energy considerations. The work done on the body of the jumper by the force F during the jump is the product of the force F and the distance c over which this force acts (see Appendix A). This work is converted to kinetic energy as the jumper is accelerated upward. At the full height of the jump H (before the jumper starts falling back to ground), the velocity of the jumper is zero. At this point, the kinetic energy is fully converted to potential energy as the center of mass of the jumper is raised to a height $(c + H)$. Therefore, from conservation of energy,

Work done on the body = Potential energy at maximum height

or

$$Fc = W(c + H) \tag{3.12}$$

From this equation the height of the jump is, as before,

$$H = \frac{(F - W)c}{W}$$

Another aspect of the vertical jump is examined in Exercise 3-1.

3.2 Effect of Gravity on the Vertical Jump

The weight of an object depends on the mass and size of the planet on which it is located. The gravitational constant of the moon, for example, is one-sixth that of the Earth; therefore, the weight of a given object on the moon is one-sixth its weight on the Earth. It is a common mistake to assume that the height to which a person can jump on the moon increases in direct proportion to the decrease in weight. This is not the case, as the following calculation will show.

From Eq. 3.11, the height of the jump on the Earth is

$$H = \frac{(F - W)\,c}{W}$$

The force F that accelerates the body upward depends on the strength of the leg muscles, and for a given person this force is the same on the moon as on the Earth. Similarly, the lowering of the center of gravity c is unchanged with location. Therefore, the height of the jump on the moon (H') is

$$H' = \frac{(F - W')c}{W'} \tag{3.13}$$

Here W' is the weight of the person on the moon (i.e., $W' = W/6$). The ratio of the jumping heights at the two locations is

$$\frac{H'}{H} = \frac{(F - W')\,W}{(F - W)\,W'} \tag{3.14}$$

If as before we assume that $F = 2W$, we find that $H'/H = 11$. That is, if a person can jump to a height of 60 cm on Earth, that same person can jump up 6.6 m on the moon. Note that the ratio $H'/H = 11$ is true only for a particular choice of F in the calculation (see Exercise 3-2).

3.3 Running High Jump

In the preceding sections, we calculated the height of a jump from a standing position and showed that the center of gravity could be raised about 60 cm. A considerably greater height can be attained by jumping from a running start. The current high-jump record is about 2.3 m. The additional height is attained by using part of the kinetic energy of the run to raise the center of gravity off the ground. Let us calculate the height attainable in a running jump if the jumper could use all his/her initial kinetic energy ($\frac{1}{2}mv^2$) to raise his/her body off the ground. If this energy were completely converted to potential energy by raising the center of gravity to a height H, then

$$MgH = \frac{1}{2}mv^2 \tag{3.15}$$

or

$$H = \frac{v^2}{2g}$$

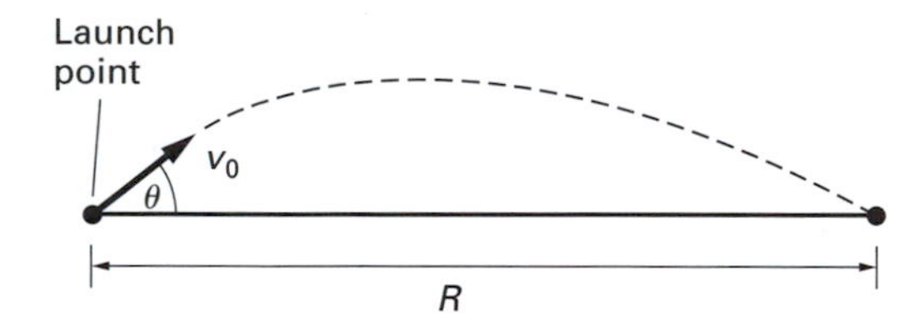

FIGURE 3.6 ▶ Projectile.

To complete our estimate, we must consider two additional factors that increase the height of the jump. First, we should add the 0.6 m, which can be produced by the legs in the final push-off. Then we must remember that the center of gravity of a person is already about 1 m above the ground. With little extra effort, the jumper can alter the position of his body so that it is horizontal at its maximum height. This maneuver adds 1 m to the height of the bar he can clear. Thus, our final estimate for the maximum height of the running high jump is

$$H = \frac{v^2}{2g} + 1.6 \text{ m} \tag{3.16}$$

The maximum short distance speed of a good runner is about 10 m/sec. At this speed, our estimate for the maximum height of the jump from Eq. 3.16 is 6.7 m. This estimate is nearly three times the high-jump record. Obviously, it is not possible for a jumper to convert all the kinetic energy of a full-speed run into potential energy.

In the unaided running high jump, only the force exerted by the feet is available to alter the direction of the running start. This limits the amount of kinetic energy that can be utilized to aid the jump. The situation is quite different in pole vaulting, where, with the aid of the pole, the jumper can in fact use most of the kinetic energy to raise his/her center of gravity. The current men's pole-vaulting record is 6.15 m (20 ft 2 in), which is remarkably close to our estimate of 6.7 m. These figures would agree even more closely had we included in our estimate the fact that the jumper must retain some forward velocity to carry him/her over the bar.

3.4 Range of a Projectile

A problem that is solved in most basic physics texts concerns a projectile launched at an angle θ and with initial velocity v_0. A solution is required for the range R, the distance at which the projectile hits the Earth (see Fig. 3.6).

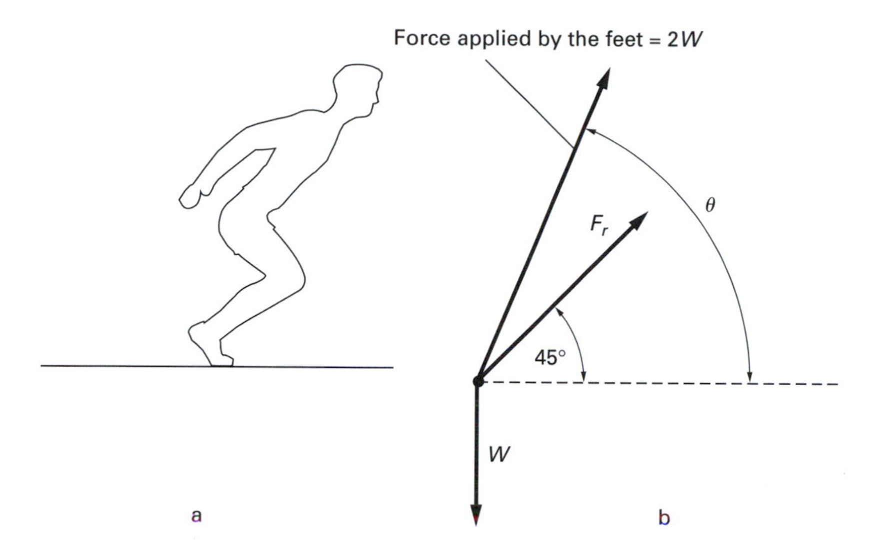

FIGURE 3.7 ▶ (a) The standing broad jump. (b) The associated forces.

It is shown that the range is

$$R = \frac{v_0^2 \sin 2\theta}{g} \tag{3.17}$$

For a given initial velocity the range is maximum when $\sin 2\theta = 1$ or $\theta = 45°$. In other words a maximum range is obtained when the projectile is launched at a 45° angle. In that case the range is

$$R_{\text{max}} = \frac{v_0^2}{g} \tag{3.18}$$

Using this result, we will estimate the distance attainable in broad jumping.

3.5 Standing Broad Jump

When the jumper projects himself into the broad jump from a stationary crouching position (Fig. 3.7), his acceleration is determined by the resultant of two forces: the downward force of gravity, which is simply equal to his weight W, and the force generated by the feet, which he can apply in any direction. In order to maximize the distance of the jump, the launching velocity and therefore also the resultant force should be directed at a 45° angle.

We will assume as before that a jumper can generate with his feet a force equal to twice the body weight. The magnitude of the resultant force (F_r) and the angle θ at which the legs must apply the force to the body are obtained from the following considerations.

The horizontal and vertical components of the resultant force (see Fig. 3.7) are, respectively,

$$\text{Horizontal component of } F_r\text{:} \quad F_r \cos 45^\circ = 2W \cos\theta \tag{3.19}$$

and

$$\text{Vertical component of } F_r\text{:} \quad F_r \sin 45^\circ = 2W \sin\theta - W \tag{3.20}$$

Here we have two equations that can be solved to yield the two unknown quantities F_r and θ (see Exercise 3-3). The magnitude of the force F_r is

$$F_r = 1.16W$$

The optimum angle θ at which the legs apply the force $2W$ is $\theta = 65.8^\circ$. We will again assume that the force that launches the jumper is applied over a distance of 60 cm, which is the extent of the crouching position. The acceleration produced by the resultant force is

$$a = \frac{F_r}{m} = \frac{1.16W}{W/g} = 1.16g$$

The launching velocity v of the jumper is therefore $v^2 = 2as$. With $s = 60$ cm, the velocity is 3.70 m/sec. The distance (R) of the standing broad jump is, from Eq. 3.18,

$$R = \frac{v^2}{g} = \frac{13.7}{9.8} = 1.4 \text{ m}$$

The range of the jump can be increased by swinging both the legs and the arms in the direction of the jump, which results in an increase in the forward momentum of the body. Other aspects of the standing broad jump are presented in Exercises 3-4 and 3-5.

3.6 Running Broad Jump

Let us assume that a jumper launches the jump from a full speed of 10 m/sec. The push-off force ($2W$) generated by the legs provides the vertical component of the launching velocity. The acceleration produced by this force is

$$a = \frac{2W}{m} = \frac{2W}{W/g} = 2g$$

If the push-off force acts on the jumper over a distance of 60 cm (the extent of the crouch) and if it is directed entirely in the vertical y direction, the vertical component of the velocity v_y during the jump is given by

$$\begin{aligned} v_y^2 &= 2as = 2 \times 2g \times 0.6 = 23.6\ \text{m}^2/\text{sec}^2 \\ v_y &= 4.87\ \text{m/sec} \end{aligned}$$

Since the horizontal component of the launching velocity v_x is the running velocity, the magnitude of the launching velocity is

$$v = \sqrt{v_x^2 + v_y^2} = 11\ \text{m/sec}$$

The launch angle θ is

$$\theta = \tan^{-1}\frac{v_y}{v_x} = \tan^{-1}\frac{4.87}{10} = 26^\circ$$

From Eq. 3.17, the range R of the jump is

$$R = \frac{v^2 \sin 2\theta}{g} = \frac{123.6 \sin 52^\circ}{g} = 9.95\ \text{m}$$

This estimate agrees closely with the current world record, which is about 9 m.

3.7 Motion through Air

We have so far neglected the effect of air resistance on the motion of objects, but we know from experience that this is not a negligible effect. When an object moves through the air, the air molecules have to be pushed out of its way. The resulting reaction force pushes back on the body and retards its motion—this is the source of fluid friction in air. We can deduce some of the properties of air friction by sticking our hand outside a moving car. Clearly, the greater the velocity with respect to the air, the larger is the resistive force. By rotating our hand, we observe that the force is greater when the palms face the direction of motion. We therefore conclude that the resistive force increases with the velocity and the surface area in the direction of motion. It has been found that the force due to air resistance F_a can be expressed approximately as

$$F_a = CAv^2 \tag{3.21}$$

where v is the velocity of the object with respect to the air, A is the area facing the direction of motion, and C is the coefficient of air friction. The coefficient

C depends somewhat on the shape of the object. In our calculations, we will use the value $C = 0.88\ \text{kg/m}^3$.

Because of air resistance, there are two forces acting on a falling body: the downward force of gravity W and the upward force of air resistance. From Newton's second law (see Appendix A), we find that the equation of motion in this case is

$$W - F_a = ma \tag{3.22}$$

When the body begins to fall, its velocity is zero and the only force acting on it is the weight; but as the body gains speed, the force of air resistance grows, and the net accelerating force on the body decreases. If the body falls from a sufficiently great height, the velocity reaches a magnitude such that the force due to air resistance is equal to the weight. Past this point, the body is no longer accelerated and continues to fall at a constant velocity, called the *terminal velocity* v_t. Because the force on the body in Eq. 3.22 is not constant, the solution of this equation cannot be obtained by simple algebraic techniques. However, the terminal velocity can be obtained without difficulty. At the terminal velocity, the downward force of gravity is canceled by the upward force of air resistance, and the net acceleration of the body is zero. That is,

$$W - F_a = 0 \tag{3.23}$$

or

$$F_a = W$$

From Eq. 3.21, the terminal velocity is therefore given by

$$v_t = \sqrt{\frac{W}{CA}} \tag{3.24}$$

From this equation the terminal velocity of a falling person with mass 70 kg and an effective area of $0.2\ \text{m}^2$ is

$$v_t = \sqrt{\frac{W}{CA}} = \sqrt{\frac{70 \times 9.8}{0.88 \times 0.2}} = 62.4\ \text{m/sec (140 mph)}$$

The terminal velocity of different-sized objects that have a similar density and shape is proportional to the square root of the linear size of the objects. This can be seen from the following argument. The weight of an object is proportional to the volume, which is in turn proportional to the cube of the linear dimension L of the object,

$$W \propto L^3$$

The area is proportional to L^2. Therefore, from Eq. 3.24, the terminal velocity is proportional to $\sqrt{L}$ as shown here:

$$v_t \propto \sqrt{\frac{W}{A}} = \sqrt{\frac{L^3}{L^2}} = \sqrt{L}$$

This result has interesting implications on the ability of animals to survive a fall. With proper training, a person can jump from a height of about 10 m without sustaining serious injury. From this height, a person hits the ground at a speed of

$$v = \sqrt{2gs} = 14 \text{ m/sec (46 ft/sec)}$$

Let us assume that this is the speed with which any animal can hit the ground without injury. At this speed, the force of air resistance on an animal the size of man is negligible compared to the weight. But a small animal is slowed down considerably by air friction at this speed. A speed of 8.6 m/sec is the terminal velocity of a 1-cm bug. (See Exercise 3-6.) Such a small creature can drop from any height without injury. Miners often encounter mice in deep coal mines but seldom rats. A simple calculation shows that a mouse can fall down a 100-m mine shaft without severe injury. However, such a fall will kill a rat.

Air friction has an important effect on the speed of falling raindrops and hailstones. Without air friction, a 1-cm diameter hailstone, for example, falling from a height of 1000 m would hit the Earth at a speed of about 140 m/sec. At such speeds the hailstone would certainly injure anyone on whom it fell. As it is, air friction slows the hailstone to a safe terminal velocity of about 8.3 m/sec. (See Exercise 3-8.)

3.8 Energy Consumed in Physical Activity

Animals do work by means of muscular movement. The energy required to perform the work is obtained from the chemical energy in the food eaten by the animal. In general, only a small fraction of the energy consumed by the muscles is converted to work. For example, in bicycling at a rate of one leg extension per second, the efficiency of the muscles is 20%. In other words only one fifth of the chemical energy consumed by the muscle is converted to work. The rest is dissipated as heat. The energy consumed per unit time during a given activity is called the metabolic rate.

Muscle efficiency depends on the type of work and on the muscles involved. In most cases, the efficiency of the muscles in converting caloric food energy to work is less than 20%. However, in our subsequent calculations we will assume a 20% muscular efficiency.

We will calculate the amount of energy consumed by a 70-kg person jumping up 60 cm for 10 minutes at a rate of one jump per second. The external mechanical work performed by the leg muscles in each jump is

$$\text{Weight} \times \text{Height of jump} = 70\text{ kg} \times 9.8 \times 0.6 = 411\text{ J}$$

The total muscle work during the 10 minutes of jumping is

$$411 \times 600\,\text{jumps} = 24.7 \times 10^4\text{ J}$$

If we assume a muscle efficiency of 20%, then in the act of jumping the body consumes

$$24.7 \times 10^4 \times 5 = 1.23 \times 10^6\text{ J} = 294 \times 10^3\text{ cal} = 294\text{ kcal}$$

This is about the energy content in two doughnuts.

In a similar vein, A. H. Cromer (see Appendix D, Bibliography) calculates the metabolic rate while running. In the calculation, it is assumed that most of the work done in running is due to the leg muscles accelerating each leg to the running speed v, and then decelerating it to 0 velocity as one leg is brought to rest and the other leg is accelerated. The work in accelerating the leg of mass m is $\frac{1}{2}mv^2$. The work done in the deceleration is also $\frac{1}{2}mv^2$. Therefore, the total amount of work done during each stride is mv^2. As is shown in Exercise 3-9, typically, a 70-kg person (leg mass 10 kg) running at 3 m/s (9-min. mile) with a muscle efficiency of 20%, and step length of 1 m, expends 1350 J/s or 1160 kcal/hr. This is in good agreement with measurements. The energy required to overcome air resistance in running is calculated in Exercise 3-10.

In connection with the energy consumption during physical activity, we should note the difference between work and muscular effort. Work is defined as the product of force and the distance over which the force acts (see Appendix A). When a person pushes against a fixed wall his/her muscles are not performing any external work because the wall does not move. Yet it is evident that considerable energy is used in the act of pushing. All the energy is expended in the body to keep the muscles balanced in the tension necessary for the act of pushing.

▶ EXERCISES ▶

3-1. Experiments show that the duration of upward acceleration in the standing vertical jump is about 0.2 sec. Calculate the power generated in a 60-cm jump by a 70-kg jumper assuming that $c = H$, as in the text.

3-2. A 70-kg astronaut is loaded so heavily with equipment that on Earth he can jump only to a height of 10 cm. How high can he jump on the Moon? (As in the text, assume that the force generated by the legs is twice the unloaded weight of the person.)

3-3. Solve Eqs. 3.19 and 3.20 for the two unknowns F_r and θ.

3-4. What is the time period in the standing broad jump during which the jumper is in the air? Assume that the conditions of the jump are as described in the text.

3-5. Consider a person on the moon who launches herself into a standing broad jump at 45°. The average force generated during launching is, as stated in the text, $F = 2W$, and the distance over which this force acts is 60 cm. Compute (a) the range of the jump; (b) the maximum height of the jump; (c) the duration of the jump.

3-6. Calculate the terminal velocity of a 1-cm bug. Assume that the density of the bug is 1 g/cm^3 and that the bug is spherical in shape with a diameter of 1 cm. Assume further that the area of the bug subject to air friction is πr^2.

3-7. Calculate the radius of a parachute that will slow a 70-kg parachutist to a terminal velocity of 14 m/sec.

3-8. Calculate the terminal velocity of a 1-cm diameter hailstone. Density of ice is 0.92 gm/cm^3. Assume that the area subject to air friction is πr^2.

3-9. Using the approach discussed in the text, calculate the energy expended per second by a person running at 3 m/sec (9-min. mile) with a muscle efficiency of 20%. Assume that the leg mass m = 10 kg and the step length is 1 m.

3-10. Compute the power necessary to overcome air resistance when running at 4.5 m/sec (6-min. mile) against a 30 km/hr wind. (Use data in the text.)

Chapter 4

Angular Motion

As was stated in Chapter 3, most natural movements of animals consist of both linear and angular motion. In this chapter, we will analyze some aspects of angular motion contained in the movement of animals. The basic equations and definitions of angular motion used in this chapter are reviewed in Appendix A.

4.1 Forces on a Curved Path

The simplest angular motion is one in which the body moves along a curved path at a constant angular velocity, as when a runner travels along a circular path or an automobile rounds a curve. The usual problem here is to calculate the centrifugal forces and determine their effect on the motion of the object.

A common problem solved in many basic physics texts requires determination of the maximum speed at which an automobile can round a curve without skidding. We will solve this problem because it leads naturally to an analysis of running. Consider a car of weight W moving on a curved level road that has a radius of curvature R. The centrifugal force F_c exerted on the moving car (see Appendix A) is

$$F_c = \frac{mv^2}{R} = \frac{Wv^2}{gR} \tag{4.1}$$

For the car to remain on the curved path, a centripetal force must be provided by the frictional force between the road and the tires. The car begins to skid on the curve when the centrifugal force is greater than the frictional force.

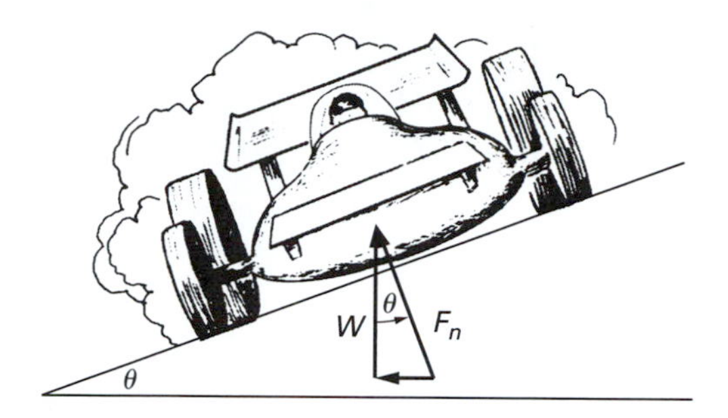

FIGURE 4.1 ▶ Banked curve.

When the car is on the verge of skidding, the centrifugal force is just equal to the frictional force; that is,

$$\frac{Wv^2}{gR} = \mu W \tag{4.2}$$

Here μ is the *coefficient of friction* between the tires and the road surface. From Eq. 4.2, the maximum velocity $v_{\max}$ without skidding is

$$v_{\max} = \sqrt{\mu g R} \tag{4.3}$$

Safe speed on a curved path may be increased by banking the road along the curve. If the road is properly banked, skidding may be prevented without recourse to frictional forces. Figure 4.1 shows a car rounding a curve banked at an angle θ. In the absence of friction, the reaction force F_n acting on the car must be perpendicular to the road surface. The vertical component of this force supports the weight of the car. That is,

$$F_n \cos\theta = W \tag{4.4}$$

To prevent skidding on a frictionless surface, the total centripetal force must be provided by the horizontal component of F_n; that is,

$$F_n \sin\theta = \frac{Wv^2}{gR} \tag{4.5}$$

where R is the radius of road curvature.

The angle θ for the road bank is obtained by taking the ratio of Eqs. 4.4 and 4.5. This yields

$$\tan\theta = \frac{v^2}{gR} \tag{4.6}$$

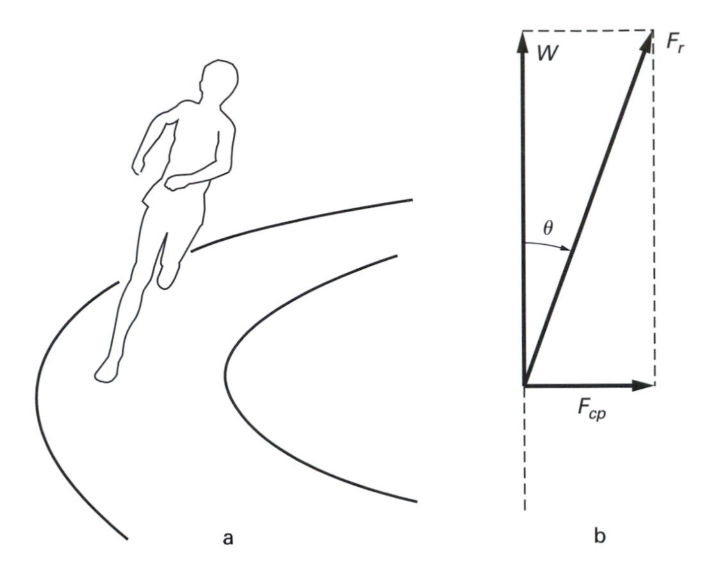

FIGURE 4.2 ▶ (a) Runner on a curved track. (b) Forces acting on the foot of the runner.

4.2 A Runner on a Curved Track

A runner on a circular track is subject to the same type of forces described in discussion of the automobile. As the runner rounds the curve, she leans toward the center of rotation (Fig. 4.2a). The reason for this position can be understood from an analysis of the forces acting on the runner. Her foot, as it makes contact with the ground, is subject to the two forces, shown in Fig. 4.2b: an upward force W, which supports her weight, and a centripetal reaction force F_{cp}, which counteracts the centrifugal force. The resultant force F_r acts on the runner at an angle θ with respect to the vertical axis.

If the runner were to round the curve remaining perpendicular to the surface, this resultant force would not pass through her center of gravity and an unbalancing torque would be applied on the runner (see Exercise 4-1). If the runner adjusts her position by leaning at an angle θ toward the center of rotation, the resultant force F_r passes through her center of gravity and the unbalancing torque is eliminated.

The angle θ is obtained from the relationships (see Fig. 4.2b)

$$F_r \sin\theta = F_{cp} = \frac{Wv^2}{gR} \tag{4.7}$$

and

$$F_r \cos\theta = W \tag{4.8}$$

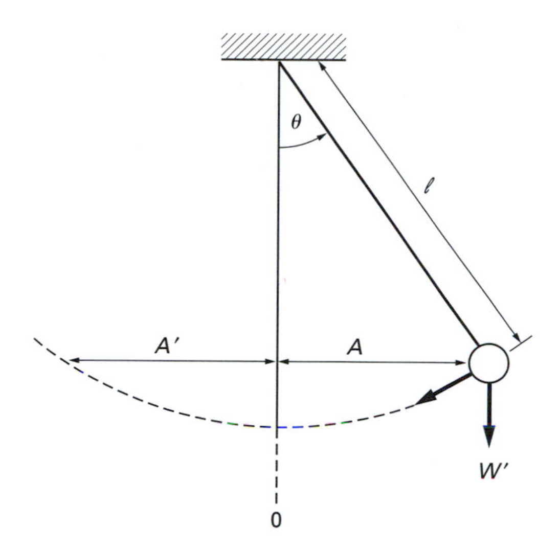

FIGURE 4.3 ▶ The simple pendulum.

Therefore

$$\tan\theta = \frac{v^2}{gR} \tag{4.9}$$

The proper angle for a speed of 6.7 m/sec (this is a 4-min. mile) on a 15-m-radius track is

$$\begin{aligned} \tan\theta &= \frac{(6.7)^2}{9.8 \times 15} = 0.305 \\ \theta &= 17^\circ \end{aligned}$$

No conscious effort is required to lean into the curve. The body automatically balances itself at the proper angle. Other aspects of centrifugal force are examined in Exercises 4-2, 4-3, and 4-4.

4.3 Pendulum

Since the limbs of animals are pivoted at the joints, the swinging motion of animals is basically angular. Many of the limb movements in walking and running can be analyzed in terms of the swinging movement of a *pendulum*.

The simple pendulum shown in Fig. 4.3 consists of a weight attached to a string, the other end of which is attached to a fixed point. If the pendulum is displaced a distance A from the center position and then released, it will swing back and forth under the force of gravity. Such a back-and-forth movement is called a *simple harmonic motion*. The number of times the pendulum swings back and forth per second is called *frequency* (f). The time for completing one cycle of the motion (i.e., from A to A' and back to A) is called the *period* T. Frequency and period are inversely related; that is, $T = 1/f$. If the angle of displacement is small, the period is given by

$$T = \frac{1}{f} = 2\pi\sqrt{\frac{\ell}{g}} \tag{4.10}$$

where g is the gravitational acceleration and ℓ is the length of the pendulum arm. Although this expression for T is derived for a small-angle swing, it is a good approximation even for a relatively wide swing. For example, when the swing is through 120° (60° in each direction), the period is only 7% longer than predicted by Eq. 4.10.

As the pendulum swings, there is continuous interchange between potential and kinetic energy. At the extreme of the swing, the pendulum is momentarily stationary. Here its energy is entirely in the form of potential energy. At this point, the pendulum, subject to acceleration due to the force of gravity, starts its return toward the center. The acceleration is tangential to the path of the swing and is at a maximum when the pendulum begins to return toward the center. The maximum tangential acceleration a_{max} at this point is given by

$$a_{\text{max}} = \frac{4\pi^2 A}{T^2} \tag{4.11}$$

As the pendulum is accelerated toward the center, its velocity increases, and the potential energy is converted to kinetic energy. The velocity of the pendulum is at its maximum when the pendulum passes the center position (0). At this point the energy is entirely in the form of kinetic energy, and the velocity (v_{max}) here is given by

$$v_{\text{max}} = \frac{2\pi A}{T} \tag{4.12}$$

4.4 Walking

Some aspects of walking can be analyzed in terms of the simple harmonic motion of a pendulum. The motion of one foot in each step can be considered as approximately a half-cycle of a simple harmonic motion (Fig. 4.4). Assume that a person walks at a rate of 120 steps/min (2 steps/sec) and that each step is 90 cm long. In the process of walking each foot rests on the ground for 0.5 sec and then swings forward 180 cm and comes to rest again 90 cm ahead

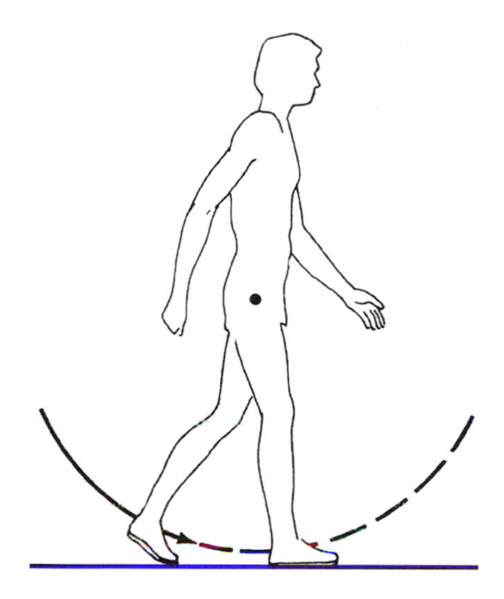

FIGURE 4.4 ▶ Walking.

of the other foot. Since the forward swing takes 0.5 sec, the full period of the harmonic motion is 1 sec. The speed of walking v is

$$v = 90 \text{ cm} \times 2 \text{ steps/sec} = 1.8 \text{ m/sec (4 mph)}$$

The maximum velocity of the swinging foot v_{max} is, from Eq. 4.12,

$$v_{\text{max}} = \frac{2\pi A}{T} = \frac{6.28 \times 90}{1} = 5.65 \text{ m/sec (12.6 mph)}$$

Thus, at its maximum velocity, the foot moves about three times faster than the body. The maximum acceleration is

$$a_{\text{max}} = \frac{4\pi^2 A}{T^2} = 35.4 \text{ m/sec}^2$$

This is 3.6 times acceleration of gravity. These formulas can also be applied to running (see Exercise 4-5).

4.5 Physical Pendulum

The simple pendulum shown in Fig. 4.3 is not an adequate representation of the swinging leg because it assumes that the total mass is located at the end of

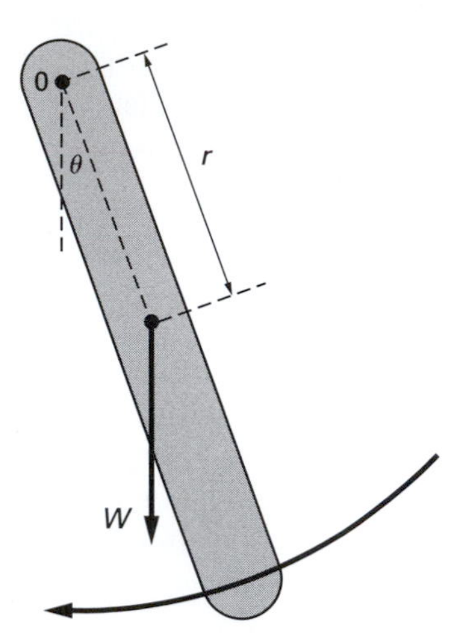

FIGURE 4.5 ▶ The physical pendulum.

the pendulum while the pendulum arm itself is weightless. A more realistic model is the physical pendulum, which takes into account the distribution of weight along the swinging object (see Fig. 4.5). It can be shown (see [6-21][1]) that under the force of gravity the period of oscillation T for a physical pendulum is

$$T = 2\pi\sqrt{\frac{I}{Wr}} \tag{4.13}$$

Here I is the moment of inertia of the pendulum around the pivot point O (see Appendix A); W is the total weight of the pendulum, and r is the distance of the center of gravity from the pivot point. (The expression for the period in Eq. 4.13 is again strictly correct only for small angular displacement.)

4.6 Speed of Walking and Running

In the analysis of walking and running, the leg may be regarded as a physical pendulum with a moment of inertia of a thin rod pivoted at one end. The moment of inertia I for the leg (see Appendix A) is, therefore,

$$I = \frac{m\ell^2}{3} = \frac{W}{g}\frac{\ell^2}{3} \tag{4.14}$$

[1] References to the bibliography are given in square brackets.

where W is the weight of the leg and ℓ is its length. If we assume that the center of mass of the leg is at its middle ($r = \frac{1}{2}\ell$), the period of oscillation is

$$T = 2\pi\sqrt{\frac{I}{Wr}} = 2\pi\sqrt{\frac{(W/g)\left(\ell^2/3\right)}{W\ell/2}} = 2\pi\sqrt{\frac{2}{3}\frac{\ell}{g}} \tag{4.15}$$

For a 90-cm-long leg, the period is 1.6 sec.

Because each step in the act of walking can be regarded as a half-swing of a simple harmonic motion, the number of steps per second is simply the inverse of the half period. In a most effortless walk, the legs swing at their natural frequency, and the time for one step is $T/2$. Walking faster or slower requires additional muscular exertion and is more tiring (see Exercise 4-6). Similar considerations apply to the swinging of the arms (see Exercise 4-7).

We can now deduce the effect of the walker's size on the speed of walking. The speed of walking is proportional to the product of the number of steps taken in a given time and the length of the step. The size of the step is in turn proportional to the length of the leg ℓ. Therefore, the speed of walking v is proportional to

$$v \propto \frac{1}{T} \times \ell \tag{4.16}$$

But because ℓ/T is proportional to $\sqrt{1/\ell}$ (see Eq. 4.15)

$$v \propto \frac{1}{\sqrt{\ell}}\ell = \sqrt{\ell} \tag{4.17}$$

Thus, the speed of the natural walk of a person increases as the square root of the length of his/her legs. The same considerations apply to all animals: The natural walk of a small animal is slower than that of a large animal.

The situation is different when a person (or any other animal) runs at full speed. Whereas in a natural walk the swing torque is produced primarily by gravity, in a fast run the torque is produced mostly by the muscles. Using some reasonable assumptions, we can show that similarly built animals can run at the same maximum speed, regardless of differences in leg size.

We assume that the length of the leg muscles is proportional to the length of the leg (ℓ) and that the area of the leg muscles is proportional to ℓ^2. The mass of the leg is proportional to ℓ^3. In other words, if one animal has a leg twice as long as that of another animal, the area of its muscle is four times as large and the mass of its leg is eight times as large.

The maximum force that a muscle can produce F_m is proportional to the area of the muscle. The maximum torque $L_{\max}$ produced by the muscle is proportional to the product of the force and the length of the leg; that is,

$$L_{\max} = F_m\ell \propto \ell^3$$

The expression in the equation for the period of oscillation is applicable for a pendulum swinging under the force of gravity. In general, the period of oscillation for a physical pendulum under the action of a torque with maximum value of $L_{\max}$ is given by

$$T = 2\pi\sqrt{\frac{I}{L_{\max}}} \tag{4.18}$$

Because the mass of the leg is proportional to ℓ^3, the moment of inertia (from Eq. 4.14) is proportional to ℓ^5. Therefore, the period of oscillation in this case is proportional to ℓ as shown

$$T \propto \sqrt{\frac{\ell^5}{\ell^3}} = \ell$$

The maximum speed of running $v_{\max}$ is again proportional to the product of the number of steps per second and the length of the step. Because the length of the step is proportional to the length of the leg, we have

$$v_{\max} \propto \frac{1}{T}\ell \propto \frac{1}{\ell} \times \ell = 1$$

This shows that the maximum speed of running is independent of the leg size, which is in accordance with observation: A fox, for example, can run at about the same speed as a horse.

Equations 4.10 and 4.15 illustrate another clearly observed aspect of running. When a person runs at a slow pace, the arms are straight as in walking. However, as the speed of running (that is the number of steps in a given interval) increases, the elbows naturally assume a bent position. In this way, the effective length (ℓ) of the pendulum decreases. This in turn increases the natural frequency of the arm, bringing it into closer synchrony with the increased frequency of steps.

4.7 Energy Expended in Running

In Chapter 3, we obtained the energy expended in running by calculating the energy needed to accelerate the leg to the speed of the run and then decelerating it to rest. Here we will use the physical pendulum as a model for the swinging leg to compute this same quantity. We will assume that in running the legs swing only at the hips. This model is, of course, not strictly correct because in running the legs swing not only at the hips but also at the knees.

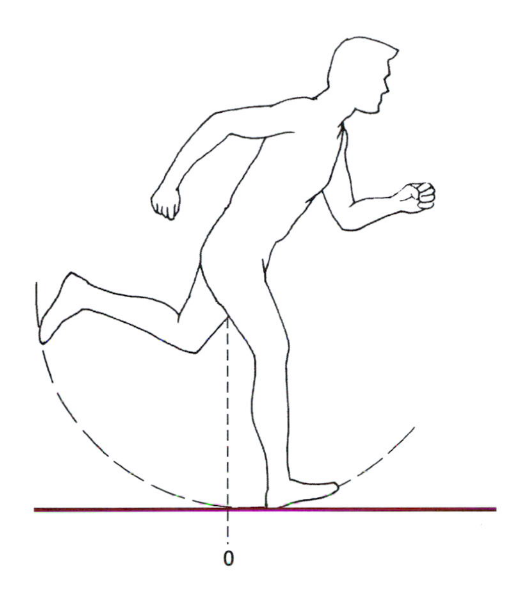

FIGURE 4.6 ▶ Running.

We will now outline a method for calculating the energy expended in swinging the legs.

During each step of the run, the leg is accelerated to a maximum angular velocity $\omega_{\max}$. In our pendulum model, this maximum angular velocity is reached as the foot swings past the vertical position 0 (see Fig. 4.6).

The rotational kinetic energy at this point is the energy provided by the leg muscles in each step of the run. This maximum rotational energy E_r (see Appendix A) is

$$E_r = \frac{1}{2} I \omega_{\max}^2$$

Here I is the moment of inertia of the leg. The angular velocity $\omega_{\max}$ is obtained as follows. From the rate of running, we can compute the period of oscillation T for the leg modeled as a pendulum. Using this value for the period, we calculate from Eq. 4.12 the maximum linear velocity $v_{\max}$ of the foot. The angular velocity (see Appendix A) is then

$$\omega_{\max} = \frac{v_{\max}}{\ell}$$

where ℓ is the length of the leg. In computing the period T, we must note that the number of steps per second each leg executes is one half of the total number of steps per second. In Exercise 4-8, it is shown that, based on the physical pendulum model for running, the amount of work done during each

TABLE 4.1 ▶ Fraction of Body Weight for Various Parts of the Body

	Fraction of body weight
Head and neck	0.07
Trunk	0.43
Upper arms	0.07
Forearms and hands	0.06
Thighs	0.23
Legs and feet	0.14
Total	1.00

From Cooper and Glassow [6-6], p. 174.

step is 1.6 mv^2. In Chapter 3, using different considerations, the amount of work done during each step was obtained as mv^2. Considering that both approaches are approximate, the agreement is certainly acceptable.

▶ EXERCISES ▶

Some of the problems in this chapter require a knowledge of the weight of human limbs. Use Table 4.1 to compute these weights.

4-1. Explain why a runner is subject to a torque if she rounds a curve maintaining a vertical position.

4-2. In the act of walking, the arms swing back and forth through an angle of 45° each second. Using the following data, calculate the maximum force on the shoulder due to the centrifugal force. The mass of the person is 70 kg, and the length of the arm is 90 cm. Assume that the total mass of the arms is located at the midpoint of the arm.

4-3. Consider the carnival ride in which the riders stand against the wall inside a large cylinder. As the cylinder rotates, the floor of the cylinder drops and the passengers are pressed against the wall by the centrifugal force. Assuming that the coefficient of friction between a rider and the cylinder wall is 0.5 and that the radius of the cylinder is 5 m, what is the minimum angular velocity of the cylinder that will hold the rider firmly against the wall?

4-4. If a person stands on a rotating pedestal with his arms loose, the arms will rise toward a horizontal position. (a) Explain the reason for this phenomenon. (b) Calculate the rotational velocity of the pedestal for the angle of the arm to be at 60° with respect to the horizontal. What is the corresponding number of revolutions per minute? Assume that the length of the arm is 90 cm.

4-5. Calculate the maximum velocity and acceleration of the foot of a runner who does a 100-m dash in 10 sec. Assume that the length of a step is 1 m and that the length of the leg is 90 cm.

4-6. What is the most effortless walking speed for a person with 90-cm-long legs if the length of each step is 90 cm?

4-7. While walking, the arms swing under the force of gravity. Compute the period of the swing. How does this period compare with the period of the leg swing? Assume arm length of 90 cm.

4-8. Using the physical pendulum model for running described in the text, derive an expression for the amount of work done during each step.

Chapter 5

Elasticity and Strength of Materials

So far we have considered the effect of forces only on the motion of a body. We will now examine the effect of forces on the shape of the body. When a force is applied to a body, the shape and size of the body change. Depending on how the force is applied, the body may be stretched, compressed, bent, or twisted. *Elasticity* is the property of a body that tends to return the body to its original shape after the force is removed. If the applied force is sufficiently large, however, the body is distorted beyond its elastic limit, and the original shape is not restored after removal of the force. A still larger force will rupture the body. We will review briefly the theory of deformation and then examine the damaging effects of forces on bones and tissue.

5.1 Longitudinal Stretch and Compression

Let us consider the effect of a stretching force F applied to a bar (Fig. 5.1). The applied force is transmitted to every part of the body, and it tends to pull the material apart. This force, however, is resisted by the cohesive force that holds the material together. The material breaks when the applied force exceeds the cohesive force. If the force in Fig. 5.1 is reversed, the bar is compressed, and its length is reduced. Similar considerations show that initially the compression is elastic, but a sufficiently large force will produce permanent deformation and then breakage.

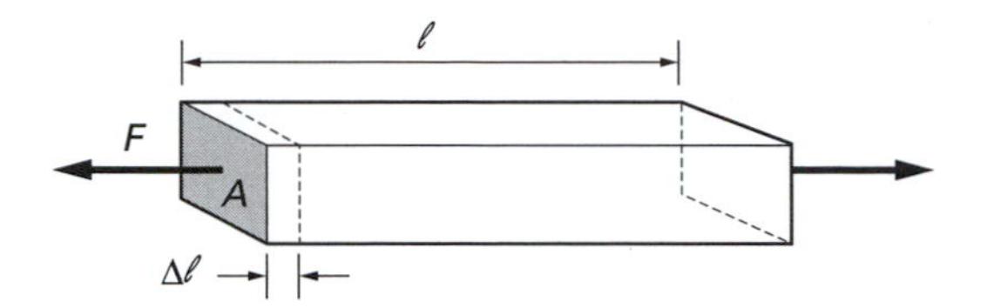

FIGURE 5.1 ▶ Stretching of a bar due to an applied force.

Stress S is the internal force per unit area acting on the material; it is defined as[1]

$$S \equiv \frac{F}{A} \tag{5.1}$$

Here F is the applied force and A is the area on which the force is applied.

The force applied to the bar in Fig. 5.1 causes the bar to elongate by an amount $\Delta\ell$. The fractional change in length $\Delta\ell/\ell$ is called the *longitudinal strain* S_t; that is,

$$S_t \equiv \frac{\Delta\ell}{\ell} \tag{5.2}$$

Here ℓ is the length of the bar and $\Delta\ell$ is the change in the length due to the applied force. If reversed, the force in Fig. 5.1 will compress the bar instead of stretching it. (Stress and strain remain defined as before.) In 1676 Robert Hooke observed that while the body remains elastic, the ratio of stress to strain is constant (Hooke's law); that is,

$$\frac{S}{S_t} = Y \tag{5.3}$$

The constant of proportionality Y is called *Young's modulus*. Young's modulus has been measured for many materials, some of which are listed in Table 5.1. The breaking or rupture strength of these materials is also shown.

5.2 A Spring

A useful analogy can be drawn between a spring and the elastic properties of a material. Consider the spring shown in Fig. 5.2.

[1]The ≡ symbol is read "defined as."

TABLE 5.1 ▶ Young's Modulus and Rupture Strength for Some Materials

Material	Young's modulus (dyn/cm^2)	Rupture strength (dyn/cm^2)
Steel	200×10^{10}	450×10^{7}
Aluminum	69×10^{10}	62×10^{7}
Bone	14×10^{10}	100×10^{7} compression
		83×10^{7} stretch
		27.5×10^{7} twist
Tendon		68.9×10^{7} stretch
Muscle		0.55×10^{7} stretch

The force F required to stretch (or compress) the spring is directly proportional to the amount of stretch; that is,

$$F = K\Delta\ell \tag{5.4}$$

The constant of proportionality K is called the *spring constant.*

A stretched (or compressed) spring contains potential energy; that is, work can be done by the stretched spring when the stretching force is removed. The energy E stored in the spring (see [6-21]) is given by

$$E = \frac{1}{2}K(\Delta\ell)^2 \tag{5.5}$$

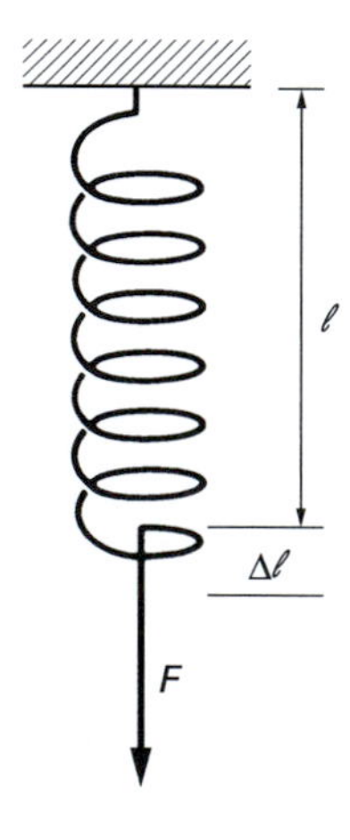

FIGURE 5.2 ▶ A stretched spring.

An elastic body under stress is analogous to a spring with a spring constant YA/ℓ. This can be seen by expanding Eq. 5.3.

$$\frac{S}{S_t} = \frac{F/A}{\Delta\ell/\ell} = Y \tag{5.6}$$

From Eq. 5.6, the force F is

$$F = \frac{YA}{\ell}\Delta\ell \tag{5.7}$$

This equation is identical to the equation for a spring with a spring constant

$$K = \frac{YA}{\ell} \tag{5.8}$$

By analogy with the spring (see Eq. 5.5), the amount of energy stored in a stretched or compressed body is

$$E = \frac{1}{2}\frac{YA}{\ell}(\Delta\ell)^2 \tag{5.9}$$

5.3 Bone Fracture: Energy Considerations

Knowledge of the maximum energy that parts of the body can safely absorb allows us to estimate the possibility of injury under various circumstances. We shall first calculate the amount of energy required to break a bone of area A and length ℓ. Assume that the bone remains elastic until fracture. Let us designate the breaking stress of the bone as S_B (see Fig. 5.3). The corresponding force F_B that will fracture the bone is, from Eq. 5.7,

$$F_B = S_B A = \frac{YA}{\ell}\Delta\ell \tag{5.10}$$

The compression $\Delta\ell$ at the breaking point is, therefore,

$$\Delta\ell = \frac{S_B\ell}{Y} \tag{5.11}$$

From Eq. 5.9, the energy stored in the compressed bone at the point of fracture is

$$E = \frac{1}{2}\frac{YA}{\ell}(\Delta\ell)^2 \tag{5.12}$$

Substituting for $\Delta\ell = S_B\ell/Y$, we obtain

$$E = \frac{1}{2}\frac{A\ell S_B^2}{Y} \tag{5.13}$$

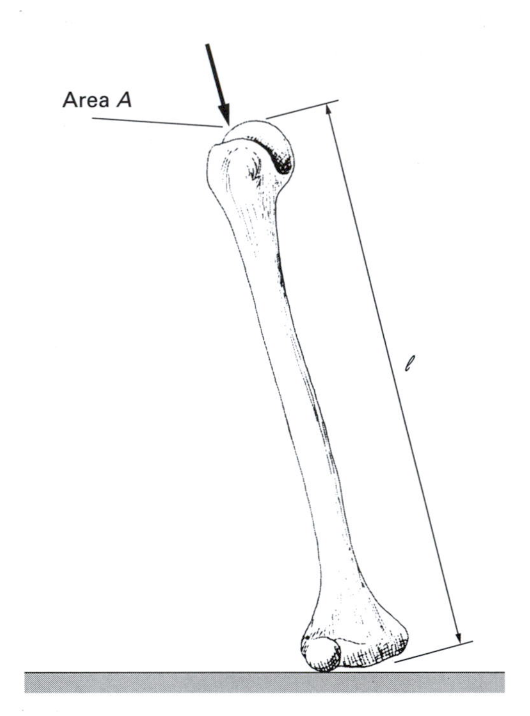

FIGURE 5.3 ▶ Compression of a bone.

As an example, consider the fracture of two leg bones that have a combined length of about 90 cm and an average area of about 6 cm^2. From Table 5.1, the breaking stress S_B is 10^9 dyn/cm^2, and Young's modulus for the bone is 14×10^{10} dyn/cm^2. The total energy absorbed by the bones of one leg at the point of compressive fracture is, from Eq. 5.13,

$$E = \frac{1}{2}\frac{6 \times 90 \times 10^{18}}{14 \times 10^{10}} = 19.25 \times 10^8 \text{ erg} = 192.5 \text{ J}$$

The combined energy in the two legs is twice this value, or 385 J. This is the amount of energy in the impact of a 70-kg person jumping from a height of 56 cm (1.8 ft), given by the product mgh. (Here m is the mass of the person, g is the gravitational acceleration, and h is the height.) If all this energy is absorbed by the leg bones, they may fracture.

It is certainly possible to jump safely from a height considerably greater than 56 cm if, on landing, the joints of the body bend and the energy of the fall is redistributed to reduce the chance of fracture. The calculation does however point out the possibility of injury in a fall from even a small height. Similar considerations can be used to calculate the possibility of bone fracture in running (see Exercise 5-1).

5.4 Impulsive Forces

In a sudden collision, a large force is exerted for a short period of time on the colliding object. The general characteristic of such a collision force as a function of time is shown in Fig. 5.4. The force starts at zero, increases to some maximum value, and then decreases to zero again. The time interval $t_2 - t_1 = \Delta t$ during which the force acts on the body is the duration of the collision. Such a short-duration force is called an *impulsive force*.

Because the collision takes place in a short period of time, it is usually difficult to determine the exact magnitude of the force during the collision. However, it is relatively easy to calculate the average value of the impulsive force F_{av}. It can be obtained simply from the relationship between force and momentum given in Appendix A; that is,

$$F_{\text{av}} = \frac{mv_f - mv_i}{\Delta t} \tag{5.14}$$

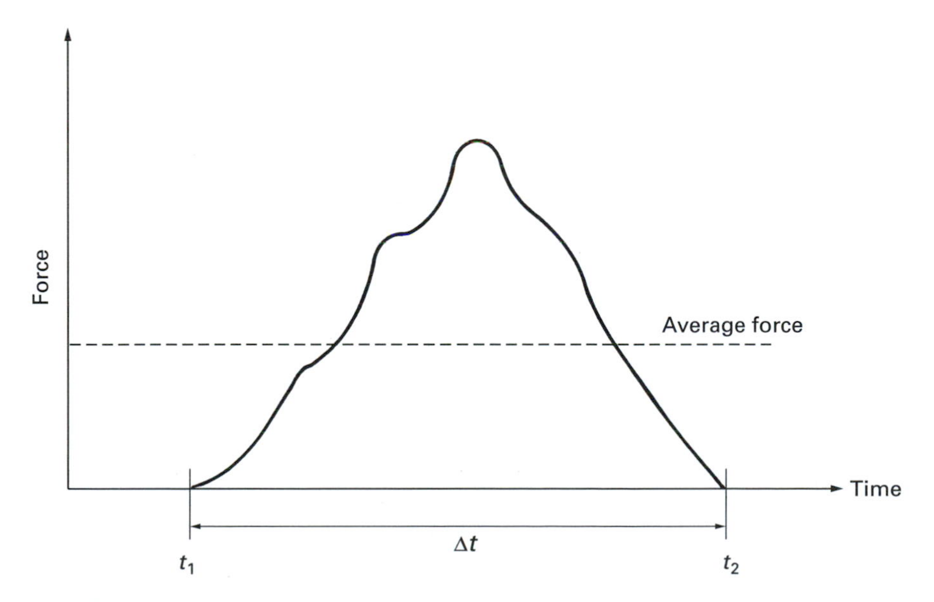

FIGURE 5.4 ▶ Impulsive force.

Here mv_i is the initial momentum of the object and mv_f is the final momentum after the collision. For example, if the duration of a collision is 6×10^{-3} sec and the change in momentum is 2 kg m/sec, the average force that acted during the collision is

$$F_{\text{av}} = \frac{2 \text{ kg m/sec}}{6 \times 10^{-3} \text{ sec}} = 3.3 \times 10^2 \text{ N}$$

Note that, for a given momentum change, the magnitude of the impulsive force is inversely proportional to the collision time; that is, the collision force is larger in a fast collision than in a slower collision.

5.5 Fracture Due to a Fall: Impulsive Force Considerations

In the preceding section, we calculated the injurious effects of collisions from energy considerations. Similar calculations can be performed using the concept of impulsive force. The magnitude of the force that causes the damage is computed from Eq. 5.14. The change in momentum due to the collision is usually easy to calculate, but the duration of the collision Δt is difficult to determine precisely. It depends on the type of collision. If the colliding objects are hard, the collision time is very short, a few milliseconds. If one of the objects is soft and yields during the collision, the duration of the collision is lengthened, and as a result the impulsive force is reduced. Thus, falling into soft sand is less damaging than falling on a hard concrete surface.

When a person falls from a height h, his/her velocity on impact with the ground, neglecting air friction (see Eq. 3.6), is

$$v = \sqrt{2gh} \tag{5.15}$$

The momentum on impact is

$$mv = m\sqrt{2gh} = W\sqrt{\frac{2h}{g}} \tag{5.16}$$

After the impact the body is at rest, and its momentum is therefore zero ($mv_f = 0$). The change in momentum is

$$mv_i - mv_f = W\sqrt{\frac{2h}{g}} \tag{5.17}$$

The average impact force, from Eq. 5.14, is

$$F = \frac{W}{\Delta t}\sqrt{\frac{2h}{g}} = \frac{m}{\Delta t}\sqrt{2gh} \tag{5.18}$$

Now comes the difficult part of the problem: Estimate of the collision duration. If the impact surface is hard, such as concrete, and if the person falls with his/her joints rigidly locked, the collision time is estimated to be about 10^{-2} sec. The collision time is considerably longer if the person bends his/her knees or falls on a soft surface.

From Table 5.1, the force per unit area that may cause a bone fracture is 10^9 dyn/cm^2. If the person falls flat on his/her heels, the area of impact may be about 2 cm^2. Therefore, the force F_B that will cause fracture is

$$F_B = 2\text{ cm}^2 \times 10^9\text{ dyn/cm}^2 = 2 \times 10^9\text{ dyn } (4.3 \times 10^3\text{ lb})$$

From Eq. 5.18, the height h of fall that will produce such an impulsive force is given by

$$h = \frac{1}{2g}\left(\frac{F\Delta t}{m}\right)^2 \tag{5.19}$$

For a man with a mass of 70 kg, the height of the jump that will generate a fracturing average impact force (assuming $\Delta t = 10^{-2}$ sec) is given by

$$h = \frac{1}{2g}\left(\frac{F\Delta t}{m}\right)^2 = \frac{1}{2 \times 980}\left(\frac{2 \times 10^9 \times 10^{-2}}{70 \times 10^3}\right) = 41.6\text{ cm } (1.37\text{ ft})$$

This is close to the result that we obtained from energy considerations. Note, however, that the assumption of a 2-cm^2 impact area is reasonable but somewhat arbitrary. The area may be smaller or larger depending on the nature of the landing; furthermore, we have assumed that the person lands with legs rigidly straight. Exercises 5-2 and 5-3 provide further examples of calculating the injurious effect of impulsive forces.

5.6 Airbags: Inflating Collision Protection Devices

The impact force may also be calculated from the distance the center of mass of the body travels during the collision under the action of the impulsive force. This is illustrated by examining the inflatable safety device used in automobiles (see Fig. 5.5). An inflatable bag is located in the dashboard of the car. In a collision, the bag expands suddenly and cushions the impact of the passenger. The forward motion of the passenger must be stopped in about 30 cm of motion if contact with the hard surfaces of the car is to be avoided. The average deceleration (see Eq. 3.6) is given by

$$a = \frac{v^2}{2s} \tag{5.20}$$

FIGURE 5.5 ▶ Inflating collision protective device.

where v is the initial velocity of the automobile (and the passenger) and s is the distance over which the deceleration occurs. The average force that produces the deceleration is

$$F = ma = \frac{mv^2}{2s} \tag{5.21}$$

where m is the mass of the passenger.

For a 70-kg person with a 30-cm allowed stopping distance, the average force is

$$F = \frac{70 \times 10^3 v^2}{2 \times 30} = 1.17 \times 10^3 \times v^2 \text{ dyn}$$

At an impact velocity of 70 km/hr (43.5 mph), the average stopping force applied to the person is 4.45×10^6 dyn. If this force is uniformly distributed over a 1000-cm^2 area of the passenger's body, the applied force per cm^2 is 4.45×10^6 dyn. This is just below the estimated strength of body tissue.

The necessary stopping force increases as the square of the velocity. At a 105-km impact speed, the average stopping force is 10^{10} dyn and the force per cm^2 is 10^7 dyn. Such a force would probably injure the passenger.

In the design of this safety system, the possibility has been considered that the bag may be triggered during normal driving. If the bag were to remain expanded, it would impede the ability of the driver to control the vehicle; therefore, the bag is designed to remain expanded for only the short time necessary to cushion the collision. (For an estimate of this period, see Exercise 5-4.)

5.7 Whiplash Injury

Neck bones are rather delicate and can be fractured by even a moderate force. Fortunately the neck muscles are relatively strong and are capable of absorbing a considerable amount of energy. If, however, the impact is sudden, as

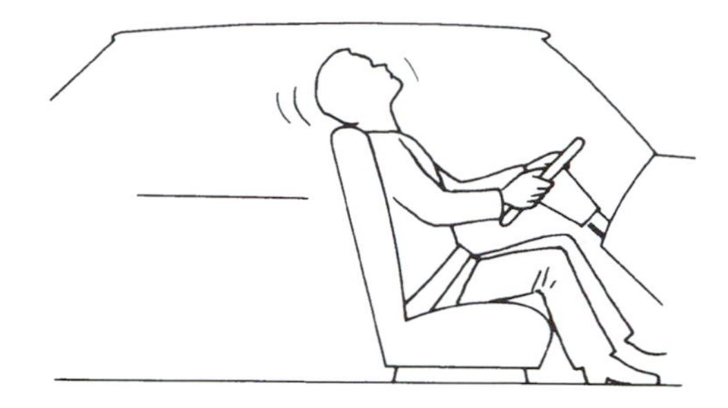

FIGURE 5.6 ▶ Whiplash.

in a rear-end collision, the body is accelerated in the forward direction by the back of the seat, and the unsupported neck is then suddenly yanked back at full speed. Here the muscles do not respond fast enough and all the energy is absorbed by the neck bones, causing the well-known whiplash injury (see Fig. 5.6). The whiplash injury is described quantitatively in Exercise 5-5.

5.8 Falling from Great Height

There have been reports of people who jumped out of airplanes with parachutes that failed to open and yet survived because they landed on soft snow. It was found in these cases that the body made about a 1-m-deep depression in the surface of the snow on impact. The credibility of these reports can be verified by calculating the impact force that acts on the body during the landing. It is shown in Exercise 5-6 that if the decelerating impact force acts over a distance of about 1 m, the average value of this force remains below the magnitude for serious injury even at the terminal falling velocity of 62.5 m/sec (140 mph).

▶ EXERCISES ▶

5-1. Assume that a 50-kg runner trips and falls on his extended hand. If the bones of one arm absorb all the kinetic energy (neglecting the energy of the fall), what is the minimum speed of the runner that will cause a fracture of the arm bone? Assume that the length of arm is 1 m and that the area of the bone is 4 cm^2.

5-2. Repeat the calculations in Exercise 5-1 using impulsive force considerations. Assume that the duration of impact is 10^{-2} sec. and the area of impact is 4 cm^2. Repeat the calculation with area of impact = 1 cm^2.

5-3. From what height can a 1-kg falling object cause fracture of the skull? Assume that the object is hard, that the area of contact with the skull is 1 cm^2, and that the duration of impact is 10^{-3} sec.

5-4. Calculate the duration of the collision between the passenger and the inflated bag of the collision protection device discussed in this chapter.

5-5. In a rear-end collision the automobile that is hit is accelerated to a velocity v in 10^{-2}/sec. What is the minimum velocity at which there is danger of neck fracture from whiplash? Use the data provided in the text, and assume that the area of the cervical vertebra is 1 cm^2 and the mass of the head is 5 kg.

5-6. Calculate the average decelerating impact force if a person falling with a terminal velocity of 62.5 m/sec is decelerated to zero velocity over a distance of 1 m. Assume that the person's mass is 70 kg and that she lands flat on her back so that the area of impact is 0.3 m^2. Is this force below the level for serious injury? (For body tissue, this is about 5×10^6 dyn/cm^2.)

5-7. A boxer punches a 50-kg bag. Just as his fist hits the bag, it travels at a speed of 7 m/sec. As a result of hitting the bag, his hand comes to a complete stop. Assuming that the moving part of his hand weighs 5 kg, calculate the rebound velocity and kinetic energy of the bag. Is kinetic energy conserved in this example? Why? (Use conservation of momentum.)

Chapter 6

Insect Flight

In this chapter, we will analyze some aspects of insect flight. In particular, we will consider the hovering flight of insects, using in our calculations many of the concepts introduced in the previous chapters. The parameters required for the computations were in most cases obtained from the literature, but some had to be estimated because they were not readily available. The size, shape, and mass of insects vary widely. We will perform our calculations for an insect with a mass of 0.1 g, which is about the size of a bee.

In general, the flight of birds and insects is a complex phenomenon. A complete discussion of flight would take into account aerodynamics as well as the changing shape of the wings at the various stages of flight. Differences in wing movements between large and small insects have only recently been demonstrated. The following discussion is highly simplified but nevertheless illustrates some of the basic physics of flight.

6.1 Hovering Flight

Many insects (and also some small birds) can beat their wings so rapidly that they are able to hover in air over a fixed spot. The wing movements in a hovering flight are complex. The wings are required to provide sideways stabilization as well as the lifting force necessary to overcome the force of gravity. The lifting force results from the downward stroke of the wings. As the wings push down on the surrounding air, the resulting reaction force of the air on the wings forces the insect up. The wings of most insects are designed so that during the upward stroke the force on the wings is small. The lifting force

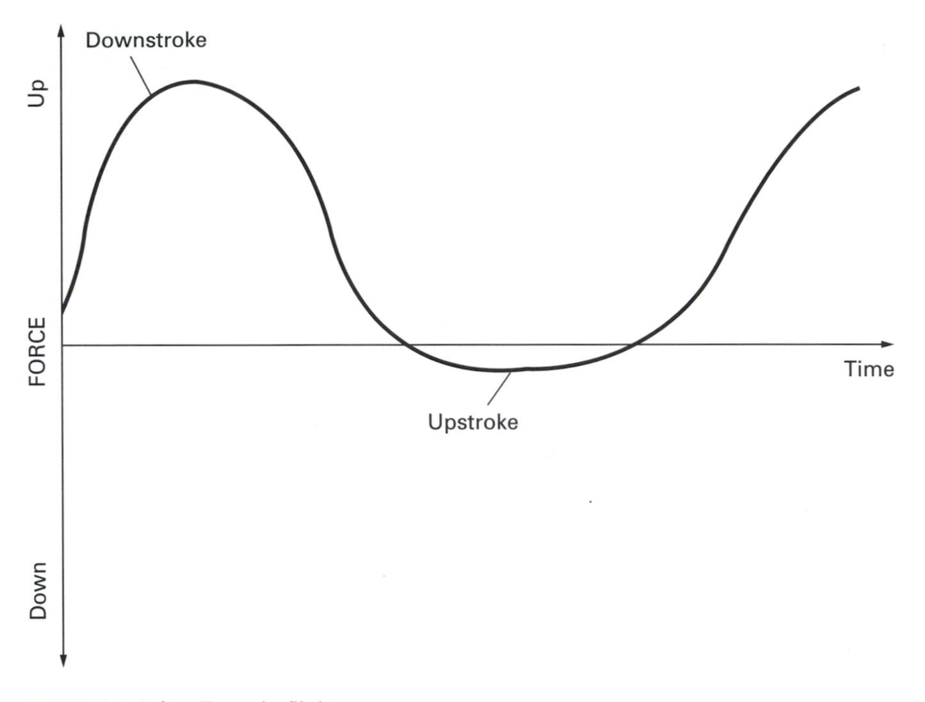

FIGURE 6.1 ▶ Force in flight.

acting on the wings during the wing movement is shown in Fig. 6.1. During the upward movement of the wings, the gravitational force causes the insect to drop. The downward wing movement then produces an upward force that restores the insect to its original position. The vertical position of the insect thus oscillates up and down at the frequency of the wingbeat.

The distance the insect falls between wingbeats depends on how rapidly its wings are beating. If the insect flaps its wings at a slow rate, the time interval during which the lifting force is zero is longer, and therefore the insect falls farther than if its wings were beating rapidly.

We can easily compute the wingbeat frequency necessary for the insect to maintain a given stability in its amplitude. To simplify the calculations, let us assume that the lifting force is at a finite constant value while the wings are moving down and that it is zero while the wings are moving up. During the time interval Δt of the upward wingbeat, the insect drops a distance h under the action of gravity. From Eq. 3.5, this distance is

$$h = \frac{g\,(\Delta t)^2}{2} \tag{6.1}$$

The upward stroke then restores the insect to its original position. Typically, it may be required that the vertical position of the insect change by no more than 0.1 mm (i.e., $h = 0.1$ mm). The maximum allowable time for free fall is then

$$\Delta t^2 = \left(\frac{2h}{g}\right)^{1/2} = \sqrt{\frac{2 \times 10^{-2}\ \text{cm}}{980\ \text{cm/sec}^2}} = 4.5 \times 10^{-3}\ \text{sec}$$

Since the up movements and the down movements of the wings are about equal in duration, the period T for a complete up-and-down wing movement is twice Δt; that is,

$$T = 2\Delta t = 9 \times 10^{-3}\ \text{sec} \tag{6.2}$$

The frequency of wingbeats f, that is, the number of wingbeats per second, is

$$f = \frac{1}{T} \tag{6.3}$$

In our example this frequency is 110 wingbeats per second. This is a typical insect wingbeat frequency, although some insects such as butterflies fly at much lower frequency, about 10 wingbeats per second (they cannot hover), and other small insects produce as many as 1000 wingbeats per second. To restore the vertical position of the insect during the downward wing stroke, the average upward force, F_{av} on the body of the insect must be equal to twice the weight of the insect (see Exercise 6-1). Note that since the upward force on the insect body is applied only for half the time, the average upward force on the insect is simply its weight.

6.2 Insect Wing Muscles

A number of different wing-muscle arrangements occur in insects. One arrangement, found in the dragonfly, is shown, highly simplified, in Fig. 6.2. The wing movement is controlled by many muscles, which are here represented by muscles A and B. The upward movement of the wings is produced by the contraction of muscle A, which depresses the upper part of the thorax and causes the attached wings to move up. While muscle A contracts, muscle B is relaxed. Note that the force produced by muscle A is applied to the wing by means of a Class 1 lever. The fulcrum here is the wing joint marked by the small circle in Fig. 6.2.

The downward wing movement is produced by the contraction of muscle B while muscle A is relaxed. Here the force is applied to the wings by means of a Class 3 lever. In our calculations, we will assume that the length of the wing is 1 cm.

The physical characteristics of insect flight muscles are not peculiar to insects. The amount of force per unit area of the muscle and the rate of muscle contraction are similar to the values measured for human muscles. Yet insect

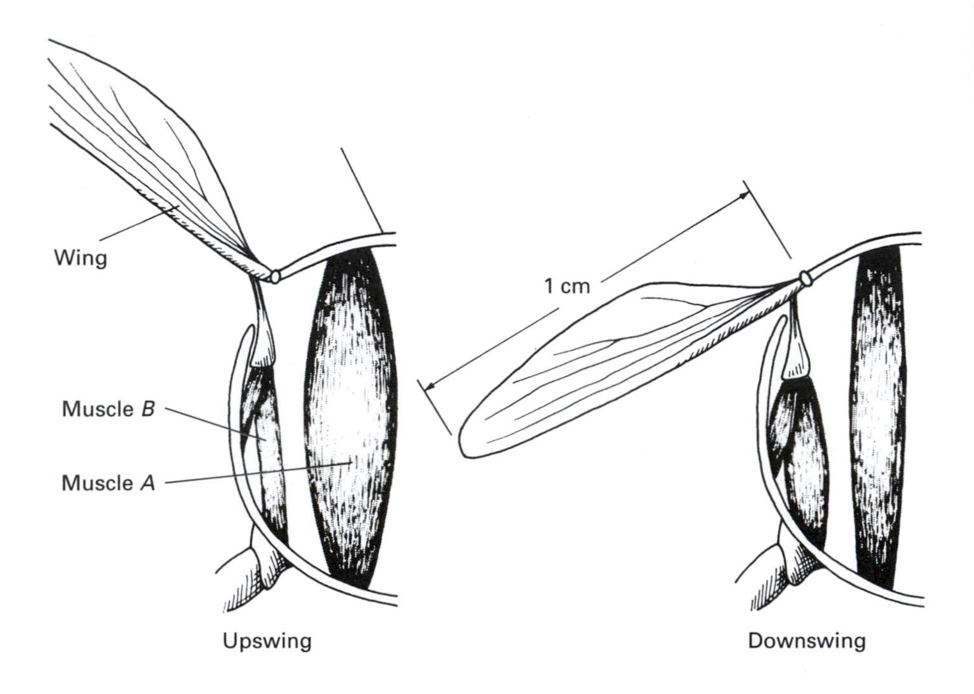

FIGURE 6.2 ▶ Wing muscles.

wing muscles are required to flap the wings at a very high rate. This is made possible by the lever arrangement of the wings. Measurements show that during a wing swing of about 70°, muscles *A* and *B* contract only about 2%. Assuming that the length of muscle *B* is 3 mm, the change in length during the muscle contraction is 0.06 mm (this is 2% of 3 mm). It can be shown that under these conditions, muscle *B* must be attached to the wing 0.052 mm from the fulcrum to achieve the required wing motion (see Exercise 6-2).

If the wingbeat frequency is 110 wingbeats per second, the period for one up-and-down motion of the wings is 9×10^{-3} sec. The downward wing movement produced by muscle *B* takes half this length of time, or 4.5×10^{-3} sec. Thus, the rate of contraction for muscle *B* is 0.06 mm divided by 4.5×10^{-3} sec, or 13 mm/sec. Such a rate of muscle contraction is commonly observed in many types of muscle tissue.

6.3 Power Required for Hovering

We will now compute the power required to maintain hovering. Let us consider again an insect with mass $m = 0.1$ g. As is shown in Exercise 6-1, the average force, F_{av}, applied by the two wings during the downward stroke is $2W$. Because the pressure applied by the wings is uniformly distributed over the total wing area, we can assume that the force generated by each wing acts through a single point at the midsection of the wings. During the downward stroke, the center of the wings traverses a vertical distance d (see Fig. 6.3). The total work done by the insect during each downward stroke is the product of force and distance; that is,

$$\text{Work} = F_{av} \times d = 2Wd \tag{6.4}$$

If the wings swing through an angle of 70°, then in our case for the insect with 1-cm-long wings d is 0.57 cm. Therefore, the work done during each stroke by the two wings is

$$\text{Work} = 2 \times 0.1 \times 980 \times 0.57 = 112 \text{ erg}$$

Let us now examine where this energy goes. In our example the mass of the insect has to be raised 0.1 mm during each downstroke. The energy E required for this task is

$$E = mgh = 0.1 \times 980 \times 10^{-2} = 0.98 \text{ erg} \tag{6.5}$$

This is a negligible fraction of the total energy expended. Clearly, most of the energy is expended in other processes. A more detailed analysis of the problem shows that the work done by the wings is converted primarily into

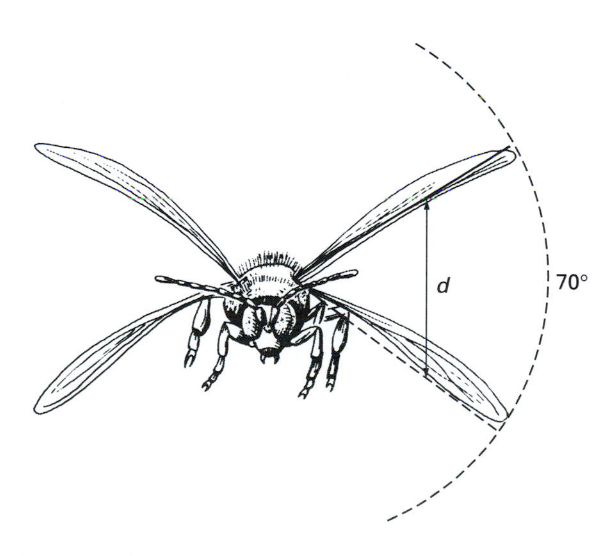

FIGURE 6.3 ▶ Insect wing motion.

kinetic energy of the air that is accelerated by the downward stroke of the wings.

Power is the amount of work done in 1 sec. Our insect makes 110 downward strokes per second; therefore, its power output P is

$$P = 112 \text{ erg} \times 110/\text{sec} = 1.23 \times 10^4 \text{ erg/sec} = 1.23 \times 10^{-3} \text{ W} \quad (6.6)$$

6.4 Kinetic Energy of Wings in Flight

In our calculation of the power used in hovering, we have neglected the kinetic energy of the moving wings. The wings of insects, light as they are, have a finite mass; therefore, as they move they possess kinetic energy. Because the wings are in rotary motion, the maximum kinetic energy during each wing stroke is

$$KE = \frac{1}{2} I \omega_{\text{max}}^2 \quad (6.7)$$

Here I is the moment of inertia of the wing and ω_{max} is the maximum angular velocity during the wing stroke. To obtain the moment of inertia for the wing, we will assume that the wing can be approximated by a thin rod pivoted at one end. The moment of inertia for the wing is then

$$I = \frac{m\ell^3}{3} \quad (6.8)$$

where ℓ is the length of the wing (1 cm in our case) and m is the mass of two wings, which may be typically 10^{-3} g. The maximum angular velocity ω_{max} can be calculated from the maximum linear velocity v_{max} at the center of the wing

$$\omega_{\text{max}} = \frac{v_{\text{max}}}{\ell/2} \quad (6.9)$$

During each stroke the center of the wings moves with an average linear velocity v_{av} given by the distance d traversed by the center of the wing divided by the duration Δt of the wing stroke. From our previous example, $d = 0.57$ cm and $\Delta t = 4.5 \times 10^{-3}$ sec. Therefore,

$$v_{\text{av}} = \frac{d}{\Delta t} = \frac{0.57}{4.5 \times 10^{-3}} = 127 \text{ cm/sec} \quad (6.10)$$

The velocity of the wings is zero both at the beginning and at the end of the wing stroke. Therefore, the maximum linear velocity is higher than the average velocity. If we assume that the velocity varies sinusoidally along the wing path, the maximum velocity is twice as high as the average velocity. Therefore, the maximum angular velocity is

$$\omega_{\text{max}} = \frac{254}{\ell/2}$$

The kinetic energy is

$$KE = \frac{1}{2} I \omega_{\text{max}}^2 = \frac{1}{2}\left(10^{-3}\frac{\ell^2}{3}\right)\left(\frac{254}{\ell/2}\right)^2 = 43 \text{ erg}$$

Since there are two wing strokes (up and down) in each cycle of the wing movement, the kinetic energy is $2 \times 43 = 86$ erg. This is about as much energy as is consumed in hovering itself.

6.5 Elasticity of Wings

As the wings are accelerated, they gain kinetic energy, which is of course provided by the muscles. When the wings are decelerated toward the end of the stroke, this energy must be dissipated. During the downstroke, the kinetic energy is dissipated by the muscles themselves and is converted into heat. (This heat is used to maintain the required body temperature of the insect.) Some insects are able to utilize the kinetic energy in the upward movement of the wings to aid in their flight. The wing joints of these insects contain a pad of elastic, rubberlike protein called *resilin* (Fig. 6.4). During the upstroke of the wing, the resilin is stretched. The kinetic energy of the wing is converted into potential energy in the stretched resilin, which stores the energy much like a spring. When the wing moves down, this energy is released and aids in the downstroke.

Using a few simplifying assumptions, we can calculate the amount of energy stored in the stretched resilin. Although the resilin is bent into a complex shape, we will assume in our calculation that it is a straight rod of area A and length ℓ. Furthermore, we will assume that throughout the stretch the resilin obeys Hooke's law. This is not strictly true as the resilin is stretched by a considerable amount and therefore both the area and Young's modulus change in the process of stretching.

The energy E stored in the stretched resilin is, from Eq. 5.9,

$$E = \frac{1}{2}\frac{YA\Delta\ell^2}{\ell} \tag{6.11}$$

Here Y is the Young's modulus for resilin, which has been measured to be 1.8×10^7 dyn/cm^2.

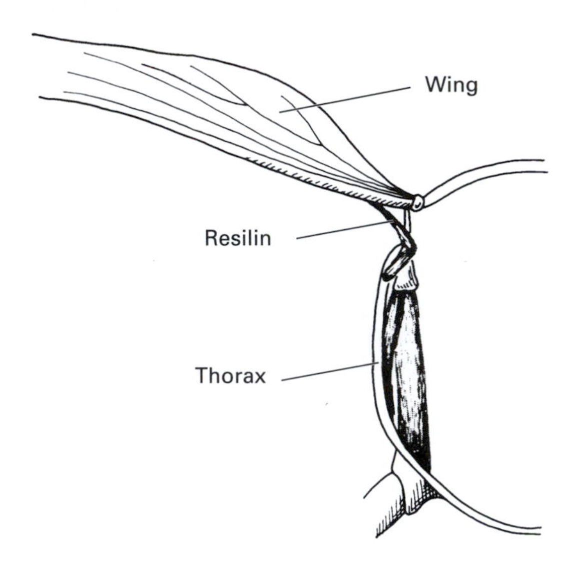

FIGURE 6.4 ► Resilin in the wing.

Typically, in an insect the size of a bee the volume of the resilin may be equivalent to a cylinder 2×10^{-2} cm long and 4×10^{-4} cm^2 in area. We will assume that the length of the resilin rod is increased by 50% when stretched. That is, $\Delta\ell$ is 10^{-2} cm. Therefore in our case the energy stored in the resilin of each wing is

$$E = \frac{1}{2}\frac{1.8 \times 10^7 \times 4 \times 10^{-4} \times 10^{-4}}{2 \times 10^{-2}} = 18 \text{ erg}$$

The stored energy in the two wings is 36 erg, which is comparable to the kinetic energy in the upstroke of the wings. Experiments show that as much as 80% of the kinetic energy of the wing may be stored in the resilin. The utilization of resilin is not restricted to wings. The hind legs of the flea, for example, also contain resilin, which stores energy for jumping (see Exercise 6-3). A further application of energy storage in resilin is examined in Exercise 6-4.

► EXERCISES ►

6-1. Compute the force on the body of the insect that must be generated during the downward wing stroke to keep the insect hovering.

6-2. Referring to the discussion in the text, compute the point of attachment to the wing of muscle B in Fig. 6.2. Assume that the muscle is perpendicular to the wing throughout the wing motion.

6-3. Assume that the shape of the resilin in each leg of the flea is equivalent to a cylinder 2×10^{-2} cm long and 10^{-4} cm^2 in area. If the change in the length of the resilin is $\Delta\ell = 10^{-2}$ cm, calculate the energy stored in the resilin. The flea weighs 0.5×10^{-3} g. How high can the flea jump utilizing only the stored energy?

6-4. Suppose that a person were equipped with resilin pads in her joints. How large would these pads have to be in order for them to store enough energy for a $\frac{1}{2}$ m jump? Assume that the pad is cubic in shape and $\Delta\ell = \frac{1}{2}\ell$.

Chapter 7

Fluids

In the previous chapters, we have examined the behavior of solids under the action of forces. In the next three chapters, we will discuss the behavior of liquids and gases, both of which play an important role in the life sciences. The differences in the physical properties of solids, liquids, and gases are explained in terms of the forces that bind the molecules. In a solid, the molecules are rigidly bound; a solid therefore has a definite shape and volume. The molecules constituting a liquid are not bound together with sufficient force to maintain a definite shape, but the binding is sufficiently strong to maintain a definite volume. A liquid adapts its shape to the vessel in which it is contained. In a gas, the molecules are not bound to each other. Therefore a gas has neither a definite shape nor a definite volume—it completely fills the vessel in which it is contained. Both gases and liquids are free to flow and are called *fluids*. Fluids and solids are governed by the same laws of mechanics, but, because of their ability to flow, fluids exhibit some phenomena not found in solid matter. In this chapter we will illustrate the properties of fluid pressure, buoyant force in liquids, and surface tension with examples from biology and zoology.

7.1 Force and Pressure in a Fluid

Solids and fluids transmit forces differently. When a force is applied to one section of a solid, this force is transmitted to the other parts of the solid with its direction unchanged. Because of a fluid's ability to flow, it transmits a force uniformly in all directions. Therefore, the pressure at any point in a fluid at rest is the same in all directions. The force exerted by a fluid at rest on any area is perpendicular to the area. A fluid in a container exerts a force on all parts of the container in contact with the fluid. A fluid also exerts a force on any object immersed in it.

The pressure in a fluid increases with depth because of the weight of the fluid above. In a fluid of constant density ρ, the difference in pressure, $P_2 - P_1$, between two points separated by a vertical distance h is

$$P_2 - P_1 = \rho g h \tag{7.1}$$

Fluid pressure is often measured in millimeters of mercury, or *torr* [after Evangelista Torricelli (1608–1674), the first person to understand the nature of atmospheric pressure]. One torr is the pressure exerted by a column of mercury that is 1 mm high. *Pascal*, abbreviated as Pa is another commonly used unit of pressure. The relationship between the torr and several of the other units used to measure pressure follows:

$$\begin{aligned} 1 \text{ torr} &= 1 \text{ mm Hg} \\ &= 13.5 \text{ mm water} \\ &= 1.33 \times 10^3 \text{ dyn/cm}^2 \\ &= 1.32 \times 10^{-3} \text{ atm} \\ &= 1.93 \times 10^{-2} \text{ psi} \\ &= 1.33 \times 10^2 \text{ Pa (N/m}^2\text{)} \end{aligned} \tag{7.2}$$

7.2 Pascal's Principle

When a force F_1 is applied on a surface of a liquid that has an area A_1, the pressure in the liquid increases by an amount P (see Fig. 7.1), given by

$$P = \frac{F_1}{A_1} \tag{7.3}$$

In an incompressible liquid, the increase in the pressure at any point is transmitted undiminished to all other points in the liquid. This is known as *Pascal's principle*. Because the pressure throughout the fluid is the same, the force F_2 acting on the area A_2 in Fig. 7.1 is

$$F_2 = PA_2 = \frac{A_2}{A_1} F_1 \tag{7.4}$$

The ratio A_2/A_1 is analogous to the mechanical advantage of a lever.

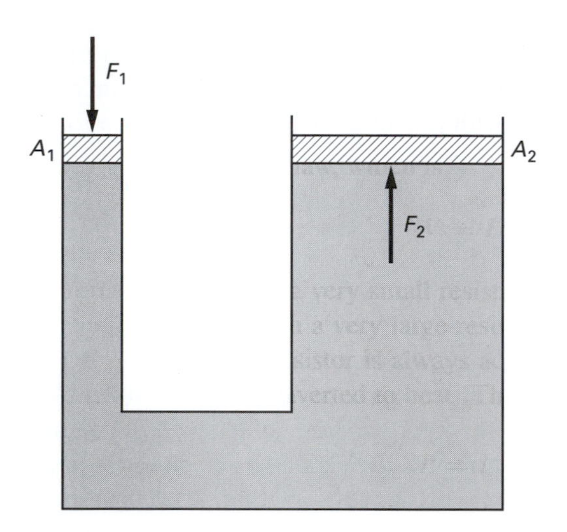

FIGURE 7.1 ▶ An illustration of Pascal's principle.

7.3 Hydrostatic Skeleton

We showed in Chapter 1 that muscles produce movement by pulling on the bones of the skeleton. There are, however, soft-bodied animals (such as the sea anemone and the earthworm) that lack a firm skeleton. Many of these animals utilize Pascal's principle to produce body motion. The structure by means of which this is done is called the *hydrostatic skeleton*.

For the purpose of understanding the movements of an animal such as a worm, we can think of the animal as consisting of a closed elastic cylinder filled with a liquid; the cylinder is its hydrostatic skeleton. The worm produces its movements with the longitudinal and circular muscles running along the walls of the cylinder (see Fig. 7.2). Because the volume of the liquid in the cylinder is constant, contraction of the circular muscles makes the worm thinner and longer. Contraction of the longitudinal muscles causes the animal to become shorter and fatter. If the longitudinal muscles contract only on one side, the animal bends toward the contracting side. By anchoring alternate ends of its body to a surface and by producing sequential longitudinal and circular contractions, the animal moves itself forward or backward. Longitudinal contraction on one side changes the direction of motion.

Let us now calculate the hydrostatic forces inside a moving worm. Consider a worm that has a radius r. Assume that the circular muscles running around its circumference are uniformly distributed along the length of the worm and that the effective area of the muscle per unit length of the worm is A_M. As the circular muscles contract, they generate a force f_M, which, along

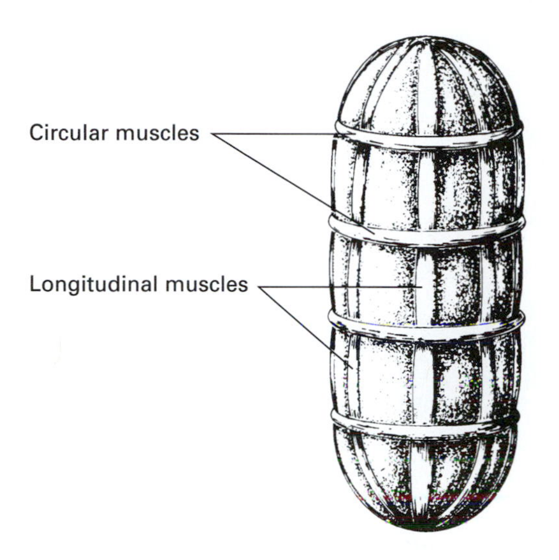

FIGURE 7.2 ▶ The hydrostatic skeleton.

each centimeter of the worm's length, is

$$f_M = SA_M \tag{7.5}$$

Here S is the force produced per unit area of the muscle. (Note that f_M is in units of force per unit length.) This force produces a pressure inside the worm. The magnitude of the pressure can be calculated with the aid of Fig. 7.3, which shows a section of the worm. The length of the section is L. If we were to cut this section in half lengthwise, as shown in Fig. 7.3, the force due to the pressure inside the cylinder would tend to push the two halves apart. This force is calculated as follows. The surface area A along the cut midsection is

$$A = L \times 2r \tag{7.6}$$

Because fluid pressure always acts perpendicular to a given surface area, the force F_P that tends to split the cylinder is

$$F_P = P \times A = P \times L \times 2r \tag{7.7}$$

Here P is the fluid pressure produced inside the worm by contraction of the circular muscles.

In equilibrium, the force F_P is balanced by the muscle forces acting along the two edges of the imaginary cut. Therefore,

$$F_P = 2f_M L$$

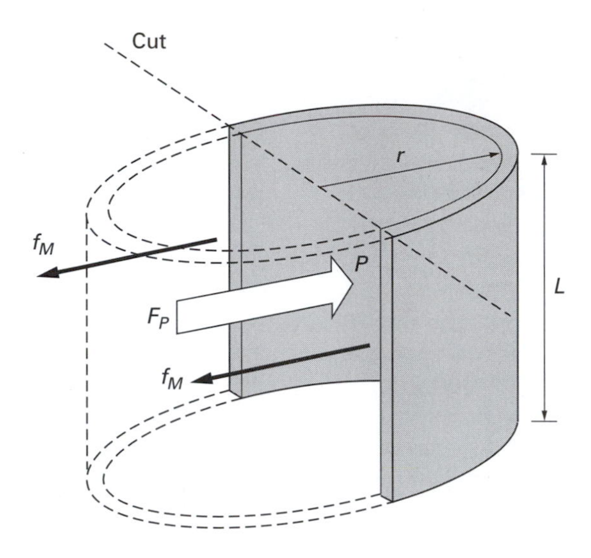

FIGURE 7.3 ▶ Calculating pressure inside a worm.

or

$$P \times L \times 2r = 2 f_M L$$

and

$$P = \frac{f_M}{r} \tag{7.8}$$

To make the calculations specific, let us assume that the radius of the worm r is 0.4 cm, the area of the circular muscles per centimeter length of the worm is $A_M = 1.5 \times 10^{-3}$ cm^2, and S, the maximum force generated per unit area of the muscle, is 7×10^6 dyn/cm^2. (This is the value we used previously for human muscles.) Therefore, the pressure inside the worm under maximum contraction of the circular muscles is

$$\begin{aligned} P &= \frac{f_M}{r} = \frac{S A_M}{r} = \frac{7 \times 10^6 \times 1.5 \times 10^{-3}}{0.4} \\ &= 2.63 \times 10^4 \text{ dyn/cm}^2 = 19.8 \text{ torr} \end{aligned}$$

This is a relatively high pressure. It can raise a column of water to a height of 26.7 cm. The force F_f in the forward direction generated by this pressure, which stretches the worm, is

$$F_f = P \times \pi r^2 = 1.32 \times 10^4 \text{ dyn}$$

The action of the longitudinal muscles can be similarly analyzed.

7.4 Archimedes' Principle

Archimedes' principle states that a body partially or wholly submerged in a fluid is buoyed upward by a force that is equal in magnitude to the weight of the displaced fluid. The derivation of this principle is found in basic physics texts. We will now use Archimedes' principle to calculate the power required to remain afloat in water and to study the buoyancy of fish.

7.5 Power Required to Remain Afloat

Whether an animal sinks or floats in water depends on its density. If its density is greater than that of water, the animal must perform work in order not to sink. We will calculate the power P required for an animal of volume V and density ρ to float with a fraction f of its volume submerged. This problem is similar to the hovering flight we discussed in Chapter 6, but our approach to the problem will be different.

Because a fraction f of the animal is submerged, the animal is buoyed up by a force F_B given by

$$F_B = gfV\rho_w \tag{7.9}$$

where ρ_w is the density of water. The force F_B is simply the weight of the displaced water.

The net downward force F_B on the animal is the difference between its weight $gV\rho$ and the buoyant force; that is,

$$F_D = gV\rho - gVf\rho_w = gV(\rho - f\rho_w) \tag{7.10}$$

To keep itself floating, the animal must produce an upward force equal to F_D. This force can be produced by pushing the limbs downward against the water. This motion accelerates the water downward and results in the upward reaction force that supports the animal.

If the area of the moving limbs is A and the final velocity of the accelerated water is v, the mass of water accelerated per unit time in the treading motion is given by (see Exercise 7-1)

$$m = Av\rho_w \tag{7.11}$$

Because the water is initially stationary, the amount of momentum imparted to the water each second is mv. (Remember that here m is the mass accelerated *per second*.)

$$\text{Momentum given to the water per second} = mv$$

This is the rate of change of momentum of the water. The force producing this change in the momentum is applied to the water by the moving limbs. The upward reaction force F_R, which supports the weight of the swimmer, is equal in magnitude to F_D and is given by

$$F_R = F_D = gV(\rho - f\rho_w) = mv \tag{7.12}$$

Substituting Eq. 7.11 for m, we obtain

$$\rho_w A v^2 = gV(\rho - f\rho_w)$$

or

$$v = \sqrt{\frac{gV(\rho - f\rho_w)}{A\rho_w}} \tag{7.13}$$

The work done by the treading limbs goes into the kinetic energy of the accelerated water. The kinetic energy given to the water each second is half the product of the mass accelerated each second and the squared final velocity of the water. This kinetic energy imparted to the water each second is the power generated by the limbs; that is,

$$KE/\text{sec} = \text{Power generated by the limbs,} \quad P = \frac{1}{2}mv^2$$

Substituting equations for m and v, we obtain (see Exercise 7-1)

$$P = \frac{1}{2}\sqrt{\frac{\left[W\left(1 - \frac{f\rho_w}{\rho}\right)\right]^3}{A\rho_w}} \tag{7.14}$$

Here W is the weight of the animal ($W = gV\rho$).

It is shown in Exercise 7-2 that a 50-kg woman expends about 7.8W to keep her nose above water. Note that, in our calculation, we have neglected the kinetic energy of the moving limbs. In Eq. 7.14 it is assumed that the density of the animal is greater than the density of water. The reverse case is examined in Exercise 7-3.

7.6 Buoyancy of Fish

The bodies of some fish contain porous bones or air-filled swim bladders that decrease their average density and allow them to float in water without an expenditure of energy. The body of the cuttlefish, for example, contains a porous bone that has a density of 0.62 g/cm^3. The rest of its body has a density of 1.067 g/cm^3. We can find the percentage of the body volume X occupied by the porous bone that makes the average density of the fish be the same as the density of sea water (1.026 g/cm^3) by using the following equation (see Exercise 7-4):

$$1.026 = \frac{0.62X + (100 - X)\,1.067}{100} \tag{7.15}$$

In this case $X = 9.2\%$.

The cuttlefish lives in the sea at a depth of about 150 m. At this depth, the pressure is 15 atm (see Exercise 7-5). The spaces in the porous bone are filled with gas at a pressure of about 1 atm. Therefore, the porous bone must be able to withstand a pressure of 14 atm. Experiments have shown that the bone can in fact survive pressures up to 24 atm.

In fish that possess swim bladders, the decrease in density is provided by the gas in the bladder. Because the density of the gas is negligible compared to the density of tissue, the volume of the swim bladder required to reduce the density of the fish is smaller than that of the porous bone. For example, to achieve the density reduction calculated in the preceding example, the volume of the bladder is only about 4% of the total volume of the fish (see Exercise 7-6).

Fish possessing porous bones or swim bladders can alter their density. The cuttlefish alters its density by injecting or withdrawing fluid from its porous bone. Fish with swim bladders alter their density by changing the amount of gas in the bladder. Another application of buoyancy is examined in Exercise 7-7.

7.7 Surface Tension

The molecules constituting a liquid exert attractive forces on each other. A molecule in the interior of the liquid is surrounded by an equal number of neighboring molecules in all directions. Therefore, the net resultant intermolecular force on an interior molecule is zero. The situation is different, however, near the surface of the liquid. Because there are no molecules above the surface, a molecule here is pulled predominantly in one direction, toward the interior of the surface. This causes the surface of a liquid to contract and behave somewhat like a stretched membrane. This contracting tendency results in a surface tension that resists an increase in the free surface of the liquid. It can be shown (see reference [7-6]) that surface tension is a force acting tangential to the surface, normal to a line of unit length on the surface (Fig. 7.4). The surface tension T of water at 25°C is 72.8 dyn/cm. The

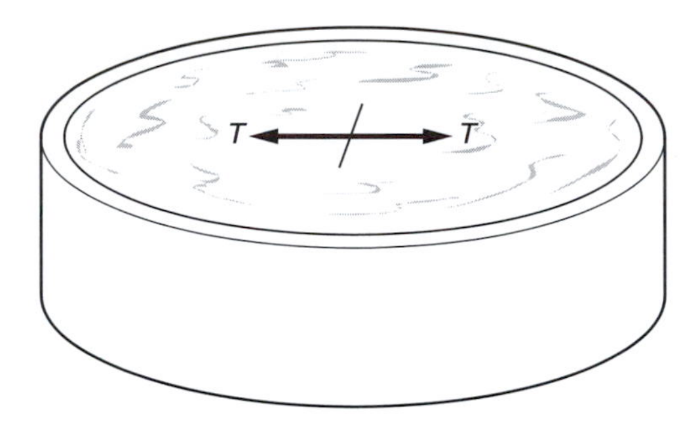

FIGURE 7.4 ▶ Surface tension.

total force F_T produced by surface tension tangential to a liquid surface of boundary length L is

$$F_T = TL \tag{7.16}$$

When a liquid is contained in a vessel, the surface molecules near the wall are attracted to the wall. This attractive force is called *adhesion*. At the same time, however, these molecules are also subject to the attractive cohesive force exerted by the liquid, which pulls the molecules in the opposite direction. If the adhesive force is greater than the cohesive force, the liquid wets the container wall, and the liquid surface near the wall is curved upward. If the opposite is the case, the liquid surface is curved downward (see Fig. 7.5). The angle θ in Fig. 7.5 is the angle between the wall and the tangent to the liquid surface at the point of contact with the wall. For a given liquid and surface material, θ is a well-defined constant. For example, the contact angle between glass and water is 25°.

If the adhesion is greater than the cohesion, a liquid in a narrow tube will rise to a specific height h (see Fig. 7.6a), which can be calculated from the

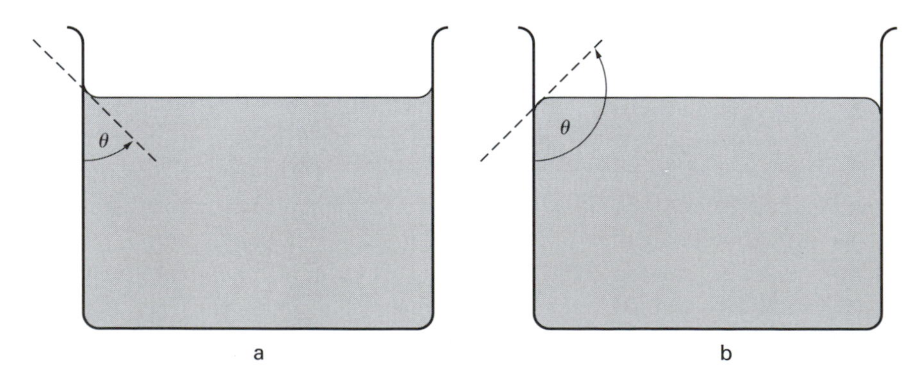

FIGURE 7.5 ▶ Angle of contact when (a) liquid wets the wall and (b) liquid does not wet the wall.

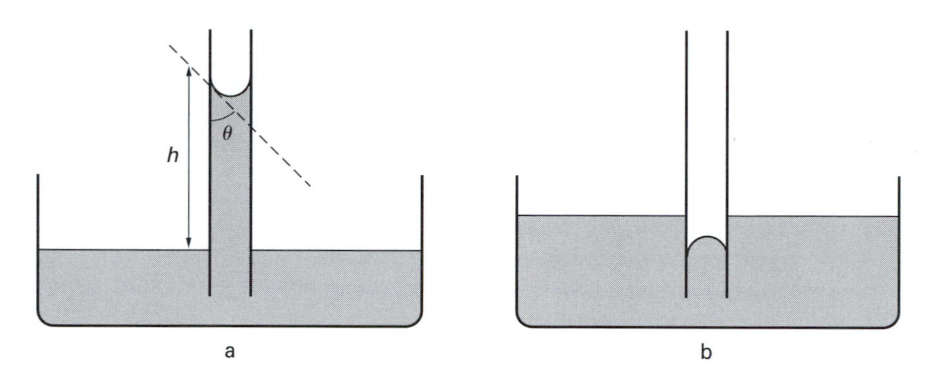

FIGURE 7.6 ▶ (a) Capillary rise. (b) Capillary depression.

following considerations. The weight W of the column of the supported liquid is

$$W = \pi R^2 h \rho g \tag{7.17}$$

where R is the radius of the column and ρ is the density of the liquid. The maximum force F_m due to the surface tension along the periphery of the liquid is

$$F_m = 2\pi RT \tag{7.18}$$

The upward component of this force supports the weight of the column of liquid (see Fig. 7.6a); that is,

$$2\pi RT \cos\theta = \pi R^2 h \rho g \tag{7.19}$$

Therefore, the height of the column is

$$h = \frac{2T\cos\theta}{R\rho g} \tag{7.20}$$

If the adhesion is smaller than the cohesion, the angle θ is greater than 90°. In this case, the height of the fluid in the tube is depressed (Fig. 7.6b). Equation 7.20 still applies, yielding a negative number for h. These effects are called *capillary action*.

Another consequence of surface tension is the tendency of liquid to assume a spherical shape. This tendency is most clearly observed in a liquid outside a container. Such an uncontained liquid forms into a sphere that can be noted in the shape of raindrops. The pressure inside the spherical liquid drop is higher than the pressure outside. The excess pressure ΔP in a liquid sphere of radius R is

$$\Delta P = \frac{2T}{R} \tag{7.21}$$

As will be shown in the following sections, the effects of surface tension are evident in many areas relevant to the life sciences.

7.8 Soil Water

Most soil is porous with narrow spaces between the small particles. These spaces act as capillaries and in part govern the motion of water through the soil. When water enters soil, it penetrates the spaces between the small particles and adheres to them. If the water did not adhere to the particles, it would run rapidly through the soil until it reached solid rock. Plant life would then be severely restricted. Because of adhesion and the resulting capillary action, a significant fraction of the water that enters the soil is retained by it. For a plant to withdraw this water, the roots must apply a negative pressure, or suction, to the moist soil. The required negative pressure may be quite high. For example, if the effective capillary radius of the soil is 10^{-3} cm, the pressure required to withdraw the water is 1.46×10^5 dyn/cm^2, or 0.144 atm (see Exercise 7-8).

The pressure required to withdraw water from the soil is called the *soil moisture tension* (SMT). The SMT depends on the grain size of the soil, its moisture content, and the material composition of the soil. The SMT is an important parameter in determining the quality of the soil. The higher the SMT, the more difficult it is for the roots to withdraw the water necessary for plant growth.

The dependence of the SMT on the grain size can be understood from the following considerations. The spaces between the particles of soil increase with the size of the grains. Because capillary action is inversely proportional to the diameter of the capillary, finely grained soil will hold water more tightly than soil of similar material with larger grains (see Fig. 7.7).

When all the pores of the soil are filled with water, the surface moisture tension is at its lowest value. In other words, under these conditions the required suction pressure produced by the plant roots to withdraw the water from the soil is the lowest. Saturated soil, however, is not the best medium for

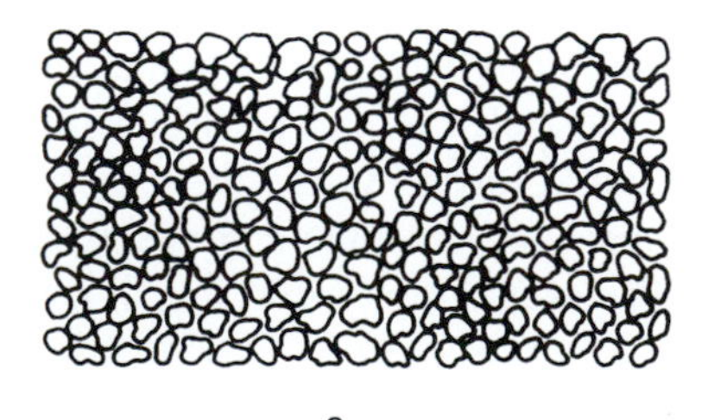

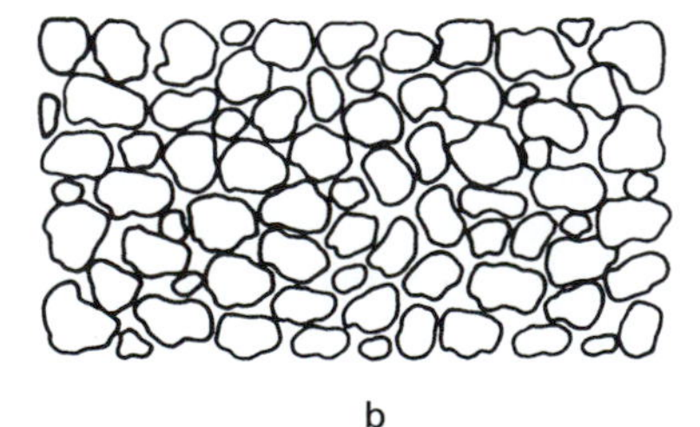

FIGURE 7.7 ▶ Fine-grained soil (a) holds water more tightly than coarse-grained soil (b).

plant growth. The roots need some air, which is absent when the soil is fully saturated with water. As the amount of water in the soil decreases, the SMT increases. In loam, for example, with a moisture content of 20% the SMT is about 0.19 atm. When the moisture content drops to 12%, the SMT increases to 0.76 atm.

The rise in SMT with decreasing moisture content can be explained in part by two effects. As the soil loses moisture, the remaining water tends to be bound into the narrower capillaries. Therefore the withdrawal of water becomes more difficult. In addition, as the moisture content decreases, sections of water become isolated and tend to form droplets. The size of these droplets may be very small. If, for example, the radius of a droplet decreases to 10^{-5} cm, the pressure required to draw the water out of the droplet is about 14.5 atm.

Capillary action also depends on the strength of adhesion, which in turn depends on the material composition of the capillary surface. For example, under similar conditions of grain size and moisture content, the SMT in clay may be ten times higher than in loam. There is a limit to the pressure that roots can produce in order to withdraw water from the soil. If the SMT increases above 15 atm, wheat, for example, cannot obtain enough water to grow. In hot dry climates where vegetation requires more water, plants may wilt even at an SMT of 2 atm. The ability of a plant to survive depends not so much on the water content as on the SMT of the soil. A plant may thrive in loam and yet wilt in a clayey soil with twice the moisture content. Other aspects of SMT are treated in Exercises 7-9 and 7-10.

7.9 Insect Locomotion on Water

About 3% of all insects are to some extent aquatic. In one way or another their lives are associated with water. Many of these insects are adapted to utilize the surface tension of water for locomotion. The surface tension of water makes it possible for some insects to stand on water and remain dry. Let us now estimate the maximum weight of an insect that can be supported by surface tension.

When the insect lands on water, the surface is depressed as shown in Fig. 7.8. The legs of such an insect, however, must not be wetted by water. A waxlike coating can provide the necessary water-repulsive property. The weight W of the insect is supported by the upward component of the surface tension; that is,

$$W = LT \sin\theta \tag{7.22}$$

where L is the combined circumference of all the insect legs in contact with the water.

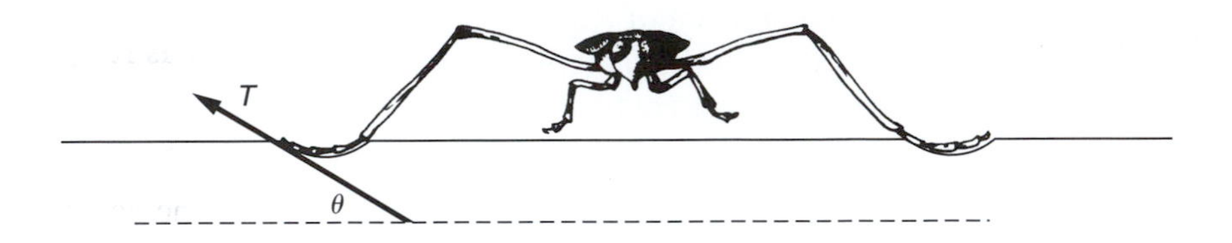

FIGURE 7.8 ▶ Insect standing on water.

To perform quantitative calculations, we must introduce some assumptions. We assume that the insect is in the shape of a cube with side dimensions ℓ. The weight of the insect of density ρ is then

$$W = \ell^3 \rho g \tag{7.23}$$

Let us further assume that the circumference of the legs in contact with water is approximately equal to the dimension of the cube; that is, from Eq. 7.23,

$$L = \ell = \left(\frac{W}{\rho g}\right)^{1/3} \tag{7.24}$$

The greatest supporting force provided by surface tension occurs at the angle $\theta = 90°$ (see Fig. 7.8). (At this point the insect is on the verge of sinking.) The maximum weight W_m that can be supported by surface tension is obtained from Eq. 7.22; that is,

$$W_m = LT = \left(\frac{W_m}{\rho g}\right)^{1/3} T$$

or

$$W_m^{2/3} = \frac{T}{(\rho g)^{1/3}} \tag{7.25}$$

If the density of the insect is 1 g/cm^3, then with $T = 72.8$ dyn/cm, the maximum weight is

$$W_m^{2/3} = \frac{72.8}{(980)^{1/3}}$$

or

$$W_m = 19.7 \text{ dyn}$$

The mass of the insect is therefore about 2×10^{-2} g. The corresponding linear size of such an insect is about 3 mm.

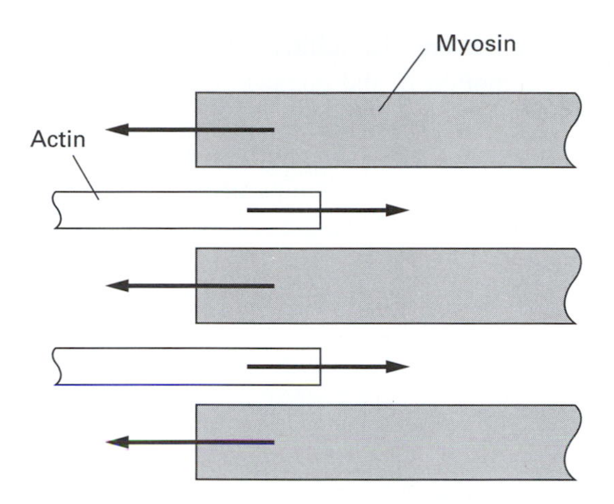

FIGURE 7.9 ▶ Contraction of muscles.

Some insects such as the beetle *Stenus* not only stand on water but also utilize surface tension for propulsion. They secrete a substance from their abdomen that reduces the surface tension behind them. As a result, they are propelled in a forward direction. Here the effect is similar to cutting a taut rubber membrane that then draws apart each section moving away from the cut (see Exercise 7-11).

7.10 Contraction of Muscles

An examination of skeletal muscles shows that they consist of smaller muscle fibers, which in turn are composed of yet smaller units called *myofibrils*. Further, examination with an electron microscope reveals that the myofibril is composed of two types of threads, one made of *myosin*, which is about 160 Å (1 Å $= 10^{-8}$ cm) in diameter, and the other made of *actin*, which has a diameter of about 50 Å. Each myosin-actin unit is about 1 mm long. The threads are aligned in a regular pattern with spaces between threads so that the threads can slide past one another, as shown in Fig. 7.9.

Muscle contraction begins with an electrical nerve impulse that results in a release of Ca^{2+} ions into the myosin-actin structure. The calcium ions in turn produce conformational changes that result in the sliding of the threads through each other, shortening the myosin-actin structure. The collective effect of this process is the contraction of the muscle.

Clearly, a force must act along the myosin-actin threads to produce such a contracting motion. The physical nature of this force is not fully understood. It has been suggested by Gamow and Ycas [7-4] that this force may be due to surface tension, which is present not only in liquids but also in jellylike materials such as tissue cells. The motion of the threads is then similar to capillary movement of a liquid. Here the movement is due to the attraction between the surfaces of the two types of thread. The surface attraction may be triggered by a release of the Ca^{2+} ions. Let us now estimate the force per square centimeter of muscle tissue that could be generated by the surface tension proposed in this model.

If the average diameter of the threads is D, the number of threads N per square centimeter of muscle is approximately

$$N = \frac{1}{\frac{\pi}{4} \times D^2} \tag{7.26}$$

The maximum pulling force F_f produced by the surface tension on each fiber is, from Eq. 7.16,

$$F_f = \pi DT \tag{7.27}$$

The total maximum force F_m due to all the fibers in a 1-cm^2 area of muscle is

$$F_m = NF_f = \frac{4T}{D} \tag{7.28}$$

The average diameter D of the muscle fibers is about 100 Å (10^{-6} cm). Therefore, the maximum contracting force that can be produced by surface tension per square centimeter of muscle area is

$$F_m = T \times 4 \times 10^6 \text{ dyn/cm}^2$$

A surface tension of 1.75 dyn/cm can account for the 7×10^6 dyn/cm^2 measured force capability of muscles. Because this is well below surface tensions commonly encountered, we can conclude that surface tension could be the source of muscle contraction. This proposed mechanism, however, should not be taken too seriously. The actual processes in muscle contraction are much more complex and cannot be reduced to a simple surface tension model (see [7-6 and 7-8]).

▶ EXERCISES ▶

7-1. Verify Eqs. 7.11 and 7.14.

7-2. With the nose above the water, about 95% of the body is submerged. Calculate the power expended by a 50-kg woman treading water in this position. Assume that the average density of the human body is about the same as water ($\rho = \rho_w = 1$ g/cm^3) and that the area A of the limbs acting on the water is about 600 cm^2.

7-3. In Eq. 7.14, it is assumed that the density of the animal is greater than the density of the fluid in which it is submerged. If the situation is reversed, the immersed animal tends to rise to the surface, and it must expend energy to keep itself below the surface. How is Eq. 7.14 modified for this case?

7-4. Derive the relationship shown in Eq. 7.15.

7-5. Calculate the pressure 150 m below the surface of the sea. The density of sea water is 1.026 g/cm^3.

7-6. Calculate the percent volume of the swim bladder in order to reduce the average density of the fish from 1.067 g/cm^3 to 1.026 g/cm^3.

7-7. The density of an animal is conveniently obtained by weighing it first in air and then immersed in a fluid. Let the weight in air and in the fluid be respectively W_1 and W_2. If the density of the fluid is ρ_1, the average density ρ_2 of the animal is

$$\rho_2 = \rho_1 \frac{W_1}{W_1 - W_2}$$

Derive this relationship.

7-8. Starting with Eq. 7.20, show that the pressure P required to withdraw the water from a capillary of radius R and contact angle θ is

$$P = \frac{2T\cos\theta}{R}$$

With the contact angle $\theta = 0°$, determine the pressure required to withdraw water from a capillary with a 10^{-3} cm radius. Assume that the surface tension $T = 72.8$ dyn/cm.

7-9. If a section of coarse-grained soil is adjacent to a finer grained soil of the same material, water will seep from the coarse-grained to the finer grained soil. Explain the reason for this.

7-10. Design an instrument to measure the SMT. (You can find a description of one such device in [7-3].)

7-11. Estimate the maximum acceleration of the insect that can be produced by reducing the surface tension as described in the text. Use size and weight data provided in the text.

Chapter 8

The Motion of Fluids

The study of fluids in motion is closely related to biology and medicine. In fact, one of the foremost workers in this field, L. M. Poiseuille (1799–1869), was a French physician whose study of moving fluids was motivated by his interest in the flow of blood through the body. In this chapter, we will review briefly the principles governing the flow of fluids and then examine the flow of blood in the circulatory system.

8.1 Bernoulli's Equation

If frictional losses are neglected, the flow of an incompressible fluid is governed by *Bernoulli's equation*, which gives the relationship between velocity, pressure, and elevation in a line of flow. Bernoulli's equation states that at any point in the channel of a flowing fluid the following relationship holds:

$$P + \rho g h + \frac{1}{2}\rho v^2 = \text{Constant} \tag{8.1}$$

Here P is the pressure in the fluid, h is the height, ρ is the density, and v is the velocity at any point in the flow channel. The first term in the equation is the potential energy per unit volume of the fluid due to the pressure in the fluid. (Note that the unit for pressure, which is dyn/cm^2, is identical to erg/cm^3, which is energy per unit volume.) The second term is the gravitational potential energy per unit volume, and the third is the kinetic energy per unit volume.

Bernoulli's equation follows from the law of energy conservation. Because the three terms in the equation represent the total energy in the fluid, in

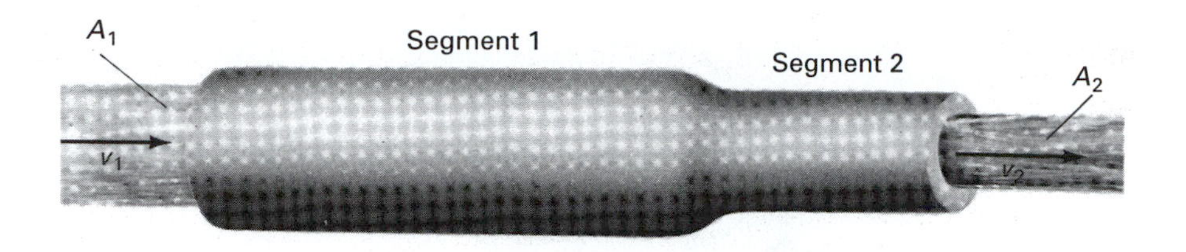

FIGURE 8.1 ▶ Flow of fluid through a pipe with two segments of different areas.

the absence of friction their sum must remain constant no matter how the flow is altered.

We will illustrate the use of Bernoulli's equation with a simple example. Consider a fluid flowing through a pipe consisting of two segments with cross-sectional areas A_1 and A_2, respectively (see Fig. 8.1). The volume of fluid flowing per second past any point in the pipe is given by the product of the fluid velocity and the area of the pipe, $A \times v$. If the fluid is incompressible, in a unit time as much fluid must flow out of the pipe as flows into it. Therefore, the rates of flow in segments 1 and 2 are equal; that is,

$$A_1 v_1 = A_2 v_2 \qquad \text{or} \qquad v_2 = \frac{A_1}{A_2} v_1 \tag{8.2}$$

In our case A_1 is larger than A_2 so we conclude that the velocity of the fluid in segment 2 is greater than in segment 1.

Bernoulli's equation states that the sum of the terms in Eq. 8.1 at any point in the flow is equal to the same constant. Therefore the relationship between the parameters P, ρ, h, and v at points 1 and 2 is

$$P_1 + \rho g h_1 + \frac{1}{2}\rho v_1^2 = P_2 + \rho g h_2 + \frac{1}{2}\rho v_2^2 \tag{8.3}$$

where the subscripts designate the parameters at the two points in the flow. Because in our case the two segments are at the same height ($h_1 = h_2$), Eq. 8.2 can be written as

$$P_1 + \frac{1}{2}\rho v_1^2 = P_2 + \frac{1}{2}\rho v_2^2 \tag{8.4}$$

Because $v_2 = (A_1/A_2)v_1$, the pressure in segment 2 is

$$P_2 = P_1 - \frac{1}{2}\rho v_1^2\left[\left(\frac{A_1}{A_2}\right)^2 - 1\right] \tag{8.5}$$

This relationship shows that while the flow velocity in segment 2 increases, the pressure in that segment decreases.

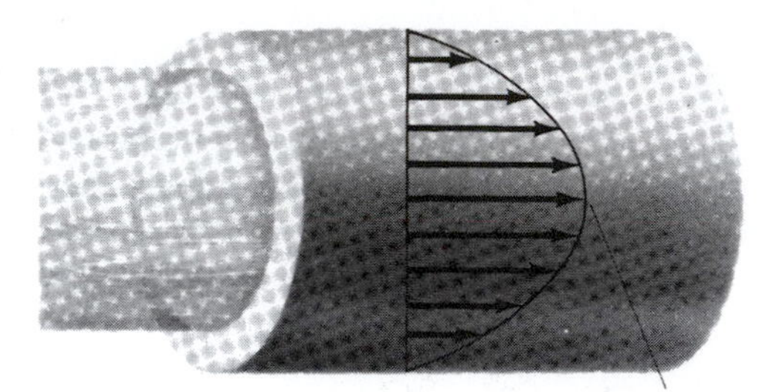

FIGURE 8.2 ▶ Laminar flow. The length of the arrows indicates the magnitude of the velocity of the fluid.

8.2 Viscosity and Poiseuille's Law

Frictionless flow is an idealization. In a real fluid, the molecules attract each other; consequently, relative motion between the fluid molecules is opposed by a frictional force, which is called *viscous friction*. Viscous friction is proportional to the velocity of flow and to the coefficient of viscosity for the given fluid. As a result of viscous friction, the velocity of a fluid flowing through a pipe varies across the pipe. The velocity is highest at the center and decreases toward the walls; at the walls of the pipe, the fluid is stationary. Such fluid flow is called *laminar*. Figure 8.2 shows the velocity profile for laminar flow in a pipe. The lengths of the arrows are proportional to the velocity across the pipe diameter.

If viscosity is taken into account, it can be shown (see reference [8-5]) that the rate of laminar flow Q through a cylindrical tube of radius R and length L is given by *Poiseuille's law*, which is

$$Q = \frac{\pi R^4 (P_1 - P_2)}{8\eta L} \text{ cm}^3/\text{sec} \tag{8.6}$$

where $P_1 - P_2$ is the difference between the fluid pressures at the two ends of the cylinder and η is the coefficient of viscosity measured in units of dyn (sec/cm^2), which is called a *poise*. The viscosities of some fluids are listed in Table 8.1. In general, viscosity is a function of temperature and increases as the fluid becomes colder.

There is a basic difference between frictionless and viscous fluid flow. A frictionless fluid will flow steadily without an external force applied to it. This fact is evident from Bernoulli's equation, which shows that if the height and velocity of the fluid remain constant, there is no pressure drop along the flow path. But Poiseuille's equation for viscous flow states that a pressure

TABLE 8.1 ▶ Viscosities of Selected Fluids

Fluid	Temperature (°C)	Viscosity (poise)
Water	20	0.01
Glycerin	20	8.3
Mercury	20	0.0155
Air	20	0.00018
Blood	37	0.04

drop always accompanies viscous fluid flow. By rearranging Eq. 8.6, we can express the pressure drop as

$$P_1 - P_2 = \frac{Q8\eta L}{\pi R^4} \tag{8.7}$$

The expression $P_1 - P_2$ is the pressure drop that accompanies the flow rate Q along a length L of the pipe. The product of the pressure drop and the area of the pipe is the force required to overcome the frictional forces that tend to retard the flow in the pipe segment. Note that for a given flow rate the pressure drop required to overcome frictional losses decreases as the fourth power of the pipe radius. Thus, even though all fluids are subject to friction, if the area of the flow is large, frictional losses and the accompanying pressure drop are small and can be neglected. In these cases, Bernoulli's equation may be used with little error.

8.3 Turbulent Flow

If the velocity of a fluid is increased past a critical point, the smooth laminar flow shown in Fig. 8.2 is disrupted. The flow becomes turbulent with eddies and whirls disrupting the laminar flow (see Fig. 8.3). In a cylindrical pipe the critical flow velocity v_c above which the flow is turbulent, is given by

$$v_c = \frac{\mathfrak{R}\eta}{\rho D} \tag{8.8}$$

Here D is the diameter of the cylinder, ρ is the density of the fluid, and η is the viscosity. The symbol $\mathfrak{R}$ is the *Reynold's number*, which for most fluids has a value between 2000 and 3000. The frictional forces in turbulent flow are greater than in laminar flow. Therefore, as the flow turns turbulent, it becomes more difficult to force a fluid through a pipe.

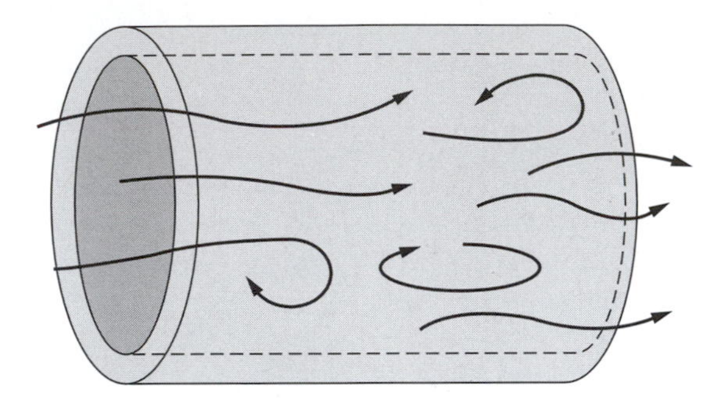

FIGURE 8.3 ▶ Turbulent fluid flow.

8.4 Circulation of the Blood

The circulation of blood through the body is often compared to a plumbing system with the heart as the pump and the veins, arteries, and capillaries as the pipes through which the blood flows. This analogy is not entirely correct. Blood is not a simple fluid; it contains cells that complicate the flow, especially when the passages become narrow. Furthermore, the veins and arteries are not rigid pipes but are elastic and alter their shape in response to the forces applied by the fluid. Still, it is possible to analyze the circulatory system with reasonable accuracy using the concepts developed for simple fluids flowing in rigid pipes.

Figure 8.4 is a drawing of the human circulatory system. The blood in the circulatory system brings oxygen, nutrients, and various other vital substances to the cells and removes the metabolic waste products from the cells. The blood is pumped through the circulatory system by the heart, and it leaves the heart through vessels called *arteries* and returns to it through *veins*.

The mammalian heart consists of two independent pumps, each made of two chambers called the *atrium* and the *ventricle*. The entrances to and exits from these chambers are controlled by valves that are arranged to maintain the flow of blood in the proper direction. Blood from all parts of the body except the lungs enters the right atrium, which contracts and forces the blood into the right ventricle. The ventricle then contracts and drives the blood through the pulmonary artery into the lungs. In its passage through the lungs, the blood releases carbon dioxide and absorbs oxygen. The blood then flows into the left atrium via the pulmonary vein. The contraction of the left atrium forces the blood into the left ventricle, which on contraction drives the oxygen-rich blood through the aorta into the arteries that lead to all parts of the body except the lungs. Thus, the right side of the heart pumps the blood through the lungs, and the left side pumps it through the rest of the body.

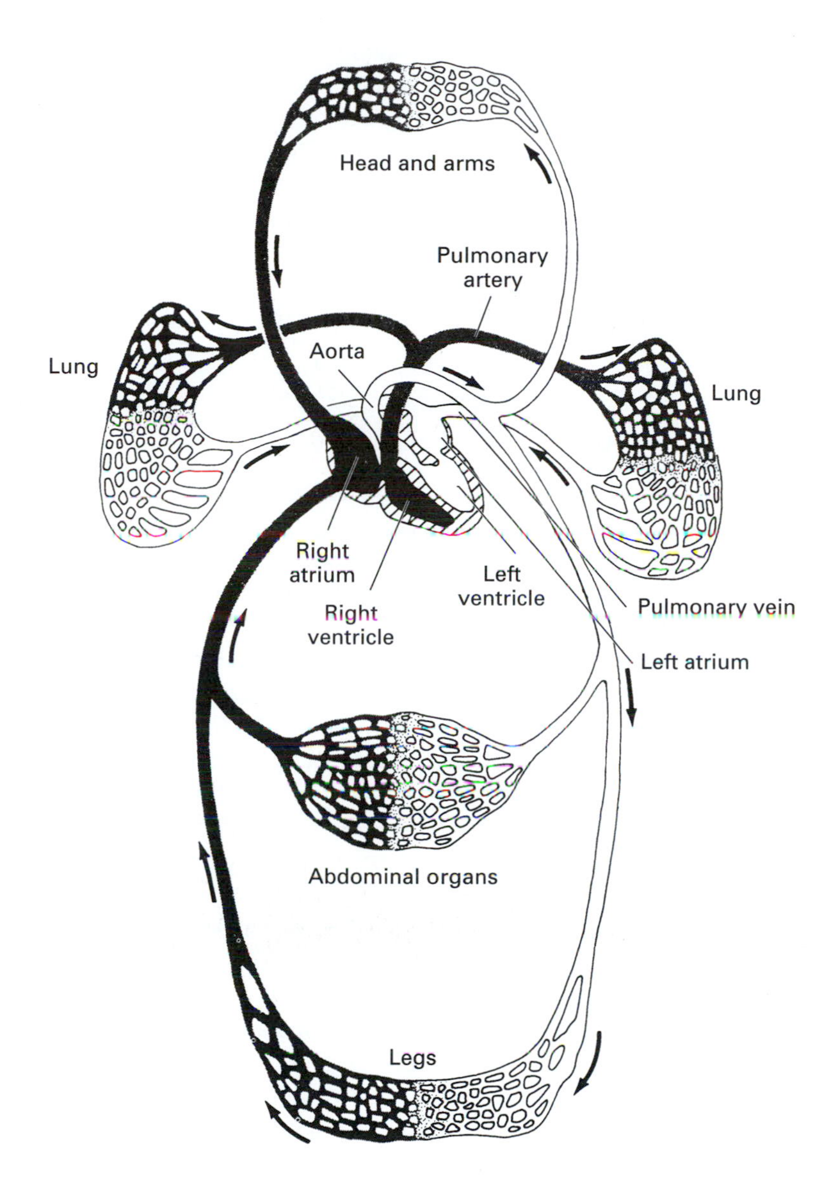

FIGURE 8.4 ▶ Schematic diagram showing various routes of the circulation.

The large artery, called the *aorta*, which carries the oxygenated blood away from the left chamber of the heart, branches into smaller arteries, which lead to the various parts of the body. These in turn branch into still smaller arteries, the smallest of which are called *arterioles*. As we will explain later, the arterioles play an important role in regulating the blood flow to specific regions in the body. The arterioles branch further into narrow capillaries that are often barely wide enough to allow the passage of single blood cells.

The capillaries are so profusely spread through the tissue that nearly all the cells in the body are close to a capillary. The exchange of gases, nutrients, and waste products between the blood and the surrounding tissue occurs by diffusion through the thin capillary walls (see Chapter 9). The capillaries join into tiny veins called *venules*, which in turn merge into larger and larger veins that lead the oxygen-depleted blood back to the right atrium of the heart.

8.5 Blood Pressure

The contraction of the heart chambers is triggered by electrical pulses that are applied simultaneously both to the left and to the right halves of the heart. First the atria contract, forcing the blood into the ventricles; then the ventricles contract, forcing the blood out of the heart. Because of the pumping action of the heart, blood enters the arteries in spurts or pulses. The maximum pressure driving the blood at the peak of the pulse is called the *systolic pressure*. The lowest blood pressure between the pulses is called the *diastolic pressure*. In a young healthy individual the systolic pressure is about 120 torr (mm Hg) and the diastolic pressure is about 80 torr. Therefore the average pressure of the pulsating blood at heart level is 100 torr.

As the blood flows through the circulatory system, its initial energy, provided by the pumping action of the heart, is dissipated by two loss mechanisms: losses associated with the *expansion and contraction* of the arterial walls and *viscous friction* associated with the blood flow. Due to these energy losses, the initial pressure fluctuations are smoothed out as the blood flows away from the heart, and the average pressure drops. By the time the blood reaches the capillaries, the flow is smooth and the blood pressure is only about 30 torr. The pressure drops still lower in the veins and is close to zero just before returning to the heart. In this final stage of the flow, the movement of blood through the veins is aided by the contraction of muscles that squeeze the blood toward the heart. One-way flow is assured by unidirectional valves in the veins.

The main arteries in the body have a relatively large radius. The radius of the aorta, for example, is about 1 cm; therefore, the pressure drop along the arteries is small. We can estimate this pressure drop using Poiseuille's law (Eq. 8.7). However, to solve the equation, we must know the rate of blood flow. The rate of blood flow Q through the body depends on the level of physical activity. At rest, the total flow rate is about 5 liter/min. During intense activity the flow rate may rise to about 25 liter/min. Exercise 8-1

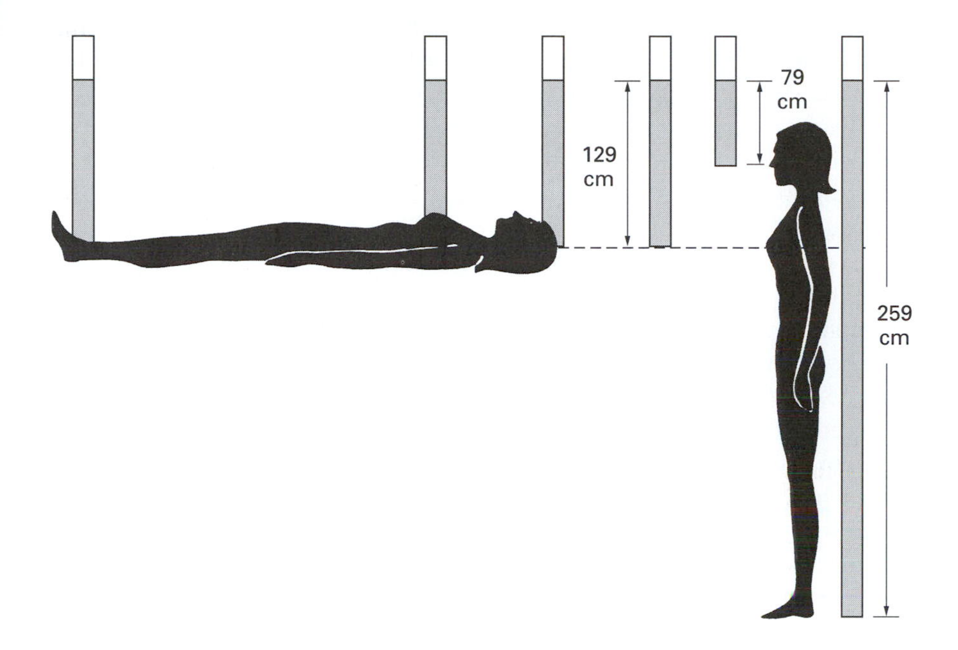

FIGURE 8.5 ▶ Blood pressure in a reclining and in an erect person.

shows that at peak flow the pressure drop per centimeter length of the aorta is only 42.5 dyn/cm^2 (3.19×10^{-2} torr), which is negligible compared to the total blood pressure.

Of course, as the aorta branches, the size of the arteries decreases, resulting in an increased resistance to flow. Although the blood flow in the narrower arteries is also reduced, the pressure drop is no longer negligible (see Exercise 8-2). The average pressure at the entrance to the arterioles is about 90 torr. Still, this is only a 10% drop from the average pressure at the heart. The flow through the arterioles is accompanied by a much larger pressure drop, about 60 torr. As a result, the pressure at the capillaries is only about 30 torr.

Since the pressure drop in the main arteries is small, when the body is horizontal, the average arterial pressure is approximately constant throughout the body. The arterial blood pressure, which is on the average 100 torr, can support a column of blood 129 cm high (see Eq. 7.1 and Exercise 8-3). This means that if a small tube were introduced into the artery, the blood in it would rise to a height of 129 cm (see Fig. 8.5).

If a person is standing erect, the blood pressure in the arteries is not uniform in the various parts of the body. The weight of the blood must be taken into account in calculating the pressure at various locations. For example, the average pressure in the artery located in the head, 50 cm above the heart (see Exercise 8-4a) is $P_{\text{head}} = P_{\text{heart}} - \rho gh = 61$ torr. In the feet, 130 cm below the heart, the arterial pressure is 200 torr (see Exercise 8-4b).

The cardiovascular system has various flow-control mechanisms that can compensate for the large arterial pressure changes that accompany shifts in the position of the body. Still, it may take a few seconds for the system to compensate. Thus, a person may feel momentarily dizzy as he/she jumps up from a prone position. This is due to the sudden decrease in the blood pressure of the brain arteries, which results in a temporary decrease of blood flow to the brain.

The same hydrostatic factors operate also in the veins, and here their effect may be more severe than in the arteries. The blood pressure in the veins is lower than in the arteries. When a person stands motionless, the blood pressure is barely adequate to force the blood from the feet back to the heart. Thus when a person sits or stands without muscular movement, blood gathers in the veins of the legs. This increases the pressure in the capillaries and may cause temporary swelling of the legs.

8.6 Control of Blood Flow

The flow of blood to specific parts of the body is controlled by the arterioles. These small vessels that receive blood from the arteries have an average diameter of about 0.1 mm. The walls of the arterioles contain smooth muscle fibers that contract when stimulated by nerve impulses and hormones. The contraction of the arterioles in one part of the body reduces the blood flow to that region and diverts it to another. Since the radius of the arterioles is small, constriction is an effective method for controlling blood flow. Poiseuille's equation shows that if the pressure drop remains constant, a 20% decrease in the radius reduces the blood flow by more than a factor of 2 (see Exercise 8-5).

8.7 Energetics of Blood Flow

For an individual at rest, the rate of blood flow is about 5 liter/min. This implies that the average velocity of the blood through the aorta is 26.5 cm/sec (see Exercise 8-6). However, the blood in the aorta does not flow continuously. It moves in spurts. During the period of flow, the velocity of the blood is about three times as high as the overall average value calculated in Exercise 8-6. Therefore, the kinetic energy per cubic centimeter of flowing blood is

$$KE = \frac{1}{2}\rho v^2 = \frac{1}{2}(1.05) \times (79.5)^2 = 3330 \text{ erg/cm}^3$$

We mentioned earlier that energy density (energy per unit volume) and pressure are measured by the same unit (i.e., 1 erg/cm^3 = 1 dyn/cm^2); therefore, they can be compared to each other. The kinetic energy of 3330 erg/cm^3 is equivalent to 2.50 torr pressure; this is small compared to the blood pressure in the aorta (which is on the average 100 torr). The kinetic energy in the smaller arteries is even less because, as the arteries branch, the overall area increases and, therefore, the flow velocity decreases. For example, when the total flow rate is 5 liter/min, the blood velocity in the capillaries is only about 0.33 mm/sec.

The kinetic energy of the blood becomes more significant as the rate of blood flow increases. For example, if during physical activity the flow rate increases to 25 liter/min, the kinetic energy of the blood is 83,300 erg/cm^3, which is equivalent to a pressure of 62.5 torr. This energy is no longer negligible compared to the blood pressure measured at rest. In healthy arteries, the increased velocity of blood flow during physical activity does not present a problem. During intense activity, the blood pressure rises to compensate for the pressure drop.

8.8 Turbulence in the Blood

Equation 8.8 shows that if the velocity of a fluid exceeds a specific critical value, the flow becomes turbulent. Through most of the circulatory system the blood flow is laminar. Only in the aorta does the flow occasionally become turbulent. Assuming a Reynold's number of 2000, the critical velocity for the onset of turbulence in the 2-cm-diameter aorta is, from Eq. 8.8,

$$V_c = \frac{\Re\eta}{\rho D} = \frac{2000 \times 0.04}{1.05 \times 2} = 38 \text{ cm/sec}$$

For the body at rest, the flow velocity in the aorta is below this value. But as the level of physical activity increases, the flow in the aorta may exceed the critical rate and become turbulent. In the other parts of the body, however, the flow remains laminar unless the passages are abnormally constricted.

Laminar flow is quiet, but turbulent flow produces noises due to vibrations of the various surrounding tissues, which indicate abnormalities in the circulatory system. These noises, called *bruit*, can be detected by a stethoscope and can help in the diagnosis of circulatory disorders.

8.9 Arteriosclerosis and Blood Flow

Arteriosclerosis is the most common of cardiovascular diseases. In the United States, an estimated 200,000 people die annually as a consequence of this disease. In arteriosclerosis, the arterial wall becomes thickened, and the artery is narrowed by deposits called *plaque*. This condition may seriously impair the functioning of the circulatory system. A 50% narrowing (*stenosis*) of the arterial area is considered moderate. Sixty to seventy percent is considered severe, and a narrowing above 80% is deemed critical. One problem caused by stenosis is made clear by Bernoulli's equation. The blood flow through the region of constriction is speeded up. If, for example, the radius of the artery is narrowed by a factor of 3, the cross-sectional area decreases by a factor of 9, which results in a nine-fold increase in velocity. In the constriction, the kinetic energy increases by 9^2, or 81. The increased kinetic energy is at the expense of the blood pressure; that is, in order to maintain the flow rate at the higher velocity, the potential energy due to pressure is converted to kinetic energy. As a result, the blood pressure in the constricted region drops. For example, if in the unobstructed artery the flow velocity is 50 cm/sec, then in the constricted region, where the area is reduced by a factor of 9, the velocity is 450 cm/sec. Correspondingly, the pressure is decreased by about 80 torr (see Exercise 8-8). Because of the low pressure inside the artery, the external pressure may actually close off the artery and block the flow of blood. When such a blockage occurs in the coronary artery, which supplies blood to the heart muscle, the heart stops functioning.

Stenosis above 80% is considered critical because at this point the blood flow usually becomes turbulent with inherently larger energy dissipation than is associated with laminar flow. As a result, the pressure drop in the situation presented earlier is even larger than calculated using Bernoulli's equation. Further, turbulent flow can damage the circulatory system because parts of the flow are directed toward the artery wall rather than parallel to it, as in laminar flow. The blood impinging on the arterial wall may dislodge some of the plaque deposit which downstream may clog a narrower part of the artery. If such clogging occurs in a cervical artery, blood flow to some part of the brain is interrupted causing an *ischemic stroke*.

There is another problem associated with arterial plaque deposit. The artery has a specific elasticity; therefore, it exhibits certain springlike properties. Specifically, in analogy with a spring, the artery has a natural frequency at which it can be readily set into vibrational motion. (See Chapter 5, Eq. 5.6.) The natural frequency of a healthy artery is in the range 1 to 2 kilohertz. Deposits of plaque cause an increase in the mass of the arterial wall and a decrease in its elasticity. As a result, the natural frequency of the artery is significantly decreased, often down to a few hundred hertz. Pulsating blood flow contains frequency components in the range of 450 hertz. The plaque-coated artery with its lowered natural frequency may now be set into resonant vibrational motion, which may dislodge plaque deposits or cause further damage to the arterial wall.

8.10 Power Produced by the Heart

The energy in the flowing blood is provided by the pumping action of the heart. We will now compute the power generated by the heart to keep the blood flowing in the circulatory system.

The power P_H produced by the heart is the product of the flow rate Q and the energy E per unit volume of the blood; that is,

$$P_H = Q\left(\frac{\text{cm}^3}{\text{sec}}\right) \times E\left(\frac{\text{erg}}{\text{cm}^2}\right) = Q \times E \text{ erg/sec} \tag{8.9}$$

At rest, when the blood flow rate is 5 liter/min, or 83.4 cm^3/sec, the kinetic energy of the blood flowing through the aorta is 3.33×10^3 erg/cm^3. (See previous section.) The energy corresponding to the systolic pressure of 120 torr is 160×10^3 erg/cm^3. The total energy is 1.63×10^5 erg/cm^3—the sum of the kinetic energy and the energy due to the fluid pressure. Therefore, the power P produced by the left ventricle of the heart is

$$P = 83.4 \times 1.63 \times 10^5 = 1.35 \times 10^7 \text{ erg/sec} = 1.35 \text{ W}$$

Exercise 8-9 shows that during intense physical activity when the flow rate increases to 25 liters/min, the peak power output of the left ventricle increases to 10.1 W.

The flow rate through the right ventricle, which pumps the blood through the lungs, is the same as the flow through the left ventricle. Here, however, the blood pressure is only one sixth the pressure in the aorta. Therefore, as shown in Exercise 8-10, the power output of the right ventricle is 0.23 W at rest and 4.5 W during intense physical activity. Thus, the total peak power output of the heart is between 1.9 and 14.6 W, depending on the intensity of the physical activity. While in fact the systolic blood pressure rises with increased blood flow, in these calculations we have assumed that it remains at 120 torr.

8.11 Measurement of Blood Pressure

The arterial blood pressure is an important indicator of the health of an individual. Both abnormally high and abnormally low blood pressures indicate some disorders in the body that require medical attention. High blood pressure, which may be caused by constrictions in the circulatory system, certainly implies that the heart is working harder than usual and that it may be endangered by the excess load. Blood pressure can be measured most directly by inserting a vertical glass tube into an artery and observing the height to which the blood rises (see Fig. 8.5). This was, in fact, the way blood pressure was first measured in 1733 by Reverend Stephen Hales, who connected a long vertical glass tube to an artery of a horse. Although sophisticated modifications of this technique are still used in special cases, this method is obviously not satisfactory for routine clinical examinations. Routine measurements of blood pressure are now most commonly performed by the cut-off method. Although this method is not as accurate as direct measurements, it is simple and in most cases adequate. In this technique, a cuff containing an inflatable balloon is placed tightly around the upper arm. The balloon is inflated with a bulb, and the pressure in the balloon is monitored by a pressure gauge. The initial pressure in the balloon is greater than the systolic pressure, and the flow of blood through the artery is therefore cut off. The observer then allows the pressure in the balloon to fall slowly by releasing some of the air. As the pressure drops, she listens with a stethoscope placed over the artery downstream from the cuff. No sound is heard until the pressure in the balloon decreases to the systolic pressure. Just below this point the blood begins to flow through the artery; however, since the artery is still partially constricted, the flow is turbulent and is accompanied by a characteristic sound. The pressure recorded at the onset of sound is the systolic blood pressure. As the pressure in the balloon drops further, the artery expands to its normal size, the flow becomes laminar, and the noise disappears. The pressure at which the sound begins to fade is taken as the diastolic pressure.

In clinical measurements, the variation of the blood pressure along the body must be considered. The cut-off blood pressure measurement is taken with the cuff placed on the arm approximately at heart level.

► EXERCISES ►

8-1. Calculate the pressure drop per centimeter length of the aorta when the blood flow rate is 25 liter/min. The radius of the aorta is about 1 cm, and the coefficient of viscosity of blood is 4×10^{-2} poise.

8-2. Compute the drop in blood pressure along a 30-cm length of artery of radius 0.5 cm. Assume that the artery carries blood at a rate of 8 liter/min.

8-3. How high a column of blood can an arterial pressure of 100 torr support? (The density of blood is 1.05 g/cm^3.)

8-4. (a) Calculate the arterial blood pressure in the head of an erect person. Assume that the head is 50 cm above the heart. (The density of blood is 1.05 g/cm^3.) (b) Compute the average arterial pressure in the legs of an erect person, 130 cm below the heart.

8-5. (a) Show that if the pressure drop remains constant, reduction of the radius of the arteriole from 0.1 to 0.08 mm decreases the blood flow by more than a factor of 2. (b) Calculate the decrease in the radius required to reduce the blood flow by 90%.

8-6. Compute the average velocity of the blood in the aorta of radius 1 cm if the flow rate is 5 liter/min.

8-7. When the rate of blood flow in the aorta is 5 liter/min, the velocity of the blood in the capillaries is about 0.33 mm/sec. If the average diameter of a capillary is 0.008 mm, calculate the number of capillaries in the circulatory system.

8-8. Compute the decrease in the blood pressure of the blood flowing through an artery the radius of which is constricted by a factor of 3. Assume that the average flow velocity in the unconstricted region is 50 cm/sec.

8-9. Using information provided in the text, calculate the power generated by the left ventricle during intense physical activity when the flow rate is 25 liter/min.

8-10. Using information provided in the text, calculate the power generated by the right ventricle during (a) restful state; blood flow 5 liter/min, and (b) intense activity; blood flow 25 liter/min.

8-11. During each heartbeat, the blood from the heart is ejected into the aorta and the pulmonary artery. Since the blood is accelerated during this part of the heartbeat, a force in the opposite direction is exerted on the rest of the body. If a person is placed on a sensitive scale (or other force-measuring device), this reaction force can be measured. An instrument based on this principle is called the *ballistocardiograph*. Discuss the type of information that might be obtained from measurements with a ballistocardiograph, and estimate the magnitude of the forces measured by this instrument.

Chapter 9

Heat and Kinetic Theory

9.1 Heat and Hotness

The sensation of hotness is certainly familiar to all of us. We know from experience that when two bodies, one hot and the other cold, are placed in an enclosure, the hotter body will cool and the colder body will heat until the degree of hotness of the two bodies is the same. Clearly something has been transferred from one body to the other to equalize their hotness. That which has been transferred from the hot body to the cold body is called *heat*. Heat may be transformed into work, and therefore it is a form of energy. Heated water, for example, can turn into steam, which can push a piston. In fact, heat can be defined as energy being transferred from a hotter body to a colder body.

In this chapter, we will discuss various properties associated with heat. We will describe the motion of atoms and molecules due to thermal energy and then discuss diffusion in connection with the functioning of cells and the respiratory system.

9.2 Kinetic Theory of Matter

To understand the present-day concept of heat, we must briefly explain the structure of matter. Matter is made of atoms and molecules, which are in continuous chaotic motion. In a gas, the atoms (or molecules) are not bound together. They move in random directions and collide frequently with one another and with the walls of the container. In addition to moving linearly, gas molecules vibrate and rotate, again in random directions. In a solid, where the atoms are bound together, the random motion is more restricted. The atoms are free only to vibrate and do so, again randomly, about some average position to which they are locked. The situation with regard to liquids is between these two extremes. Here the molecules can vibrate, but they also have some freedom to move and to rotate.

Because of their motion, the moving particles in a material possess kinetic energy. This energy of motion inside materials is called *internal energy*, and the motion itself is called *thermal motion*. What we have so far qualitatively called the hotness of a body is a measure of the internal energy; that is, in hotter bodies, the random motion of atoms and molecules is faster than in colder bodies. Therefore, the hotter an object, the greater is its internal energy. The physical sensation of hotness is the effect of this random atomic and molecular motion on the sensory mechanism. *Temperature* is a quantitative measure of hotness. The internal energy of matter is proportional to its temperature.

Using these concepts, it is possible to derive the equations that describe the behavior of matter as a function of temperature. Gases are the simplest to analyze. The theory considers a gas made of small particles (atoms or molecules) which are in continuous random motion. Each particle travels in a straight line until it collides with another particle or with the walls of the container. After a collision, the direction and speed of the particle is changed randomly. In this way kinetic energy is exchanged among the particles.

The colliding particles exchange energy not only among themselves but also with the wall of the container (Fig. 9.1). For example, if initially the walls of the container are hotter than the gas, the particles colliding with the wall on the average pick up energy from the vibrating molecules in the wall. As a result of the wall collisions, the gas is heated until it is as hot as the walls. After that, there is no net exchange of energy between the walls and the gas. This is an equilibrium situation in which, on the average, as much energy is delivered to the wall by the gas particles as is picked up from it.

The speed and corresponding kinetic energy of the individual particles in a gas vary over a wide range. Still it is possible to compute an average kinetic energy for the particles by adding the kinetic energy of all the individual particles in the container and dividing by the total number of particles (for details, see [11-7]). Many of the properties of a gas can be simply derived by assuming that each particle has this same average energy.

The internal energy in an ideal gas is in the form of kinetic energy,[1] and therefore the average kinetic energy $\left[\left(\frac{1}{2}mv^2\right)_{\text{av}}\right]$ is proportional to the temperature. The proportionality can be changed to an equality by multiplying the temperature T by a suitable constant which relates the temperature to the internal energy. The constant is designated by the symbol k, which is called *Boltzmann constant*. For historical reasons, Boltzmann constant has been so

[1]The simple theory neglects the vibrational and rotational energy of the molecules.

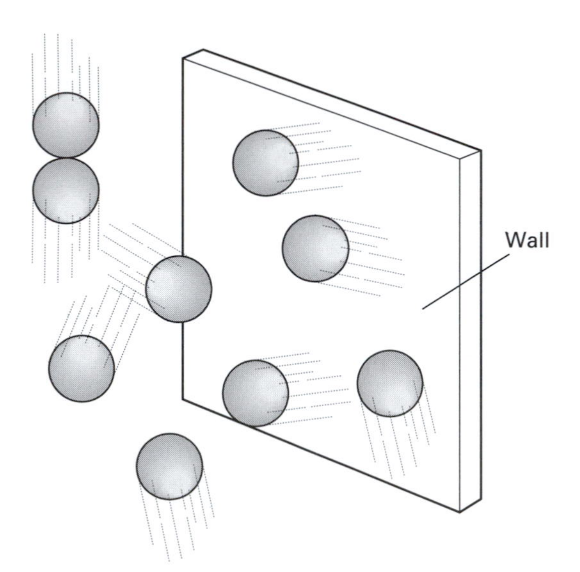

FIGURE 9.1 ▶ Collisions in a gas.

defined that it has to be multiplied by a factor of $\frac{3}{2}$ to relate temperature to the average kinetic energy of a molecule; thus,

$$\left(\frac{1}{2}mv^2\right)_{\text{av}} = \frac{3}{2}kT \tag{9.1}$$

The temperature in this equation is measured on the absolute temperature scale in degrees Kelvin. The size of the degree division on the absolute scale is equal to the Celsius, or centigrade, degree, but the absolute scale is transposed so that 0 °C = 273.15 K. Since our calculations are carried only to three significant figures, we will use simply 0 °C = 273° K. The value of Boltzmann constant is

$$k = 1.38 \times 10^{-23} \text{ J/molecule K}$$

The velocity defined by Eq. 9.1 is called *thermal velocity*.

Each time a molecule collides with the wall, momentum is transferred to the wall. The change in momentum per unit time is a force. The pressure exerted by a gas on the walls of its container is due to the numerous collisions of the gas molecules with the container. The following relationship between pressure P, volume V, and temperature is derived in most basic physics texts:

$$PV = NkT \tag{9.2}$$

Here N is the total number of gas molecules in the container of volume V, and the temperature is again measured on the absolute scale.

TABLE 9.1 ► Specific Heat for Some Substances

Substance	Specific heat (cal/g °C)
Water	1
Ice	0.480
Average for human body	0.83
Soil	0.2 to 0.8, depending on water content
Aluminum	0.214
Protein	0.4

In a closed container, the total number of particles N is fixed; therefore, if the temperature is kept unchanged, the product of pressure and volume is a constant. This is known as *Boyle's law*. (See Exercises 9-1 and 9-2.)

9.3 Definitions

9.3.1 Unit of Heat

As discussed in Appendix A, heat is measured in *calories*. One calorie (cal) is the amount of heat required to raise the temperature of 1 g of water by 1 C°.[2] Actually, because this value depends somewhat on the initial temperature of the water, the calorie is defined as the heat required to raise the temperature of 1 g of water from 14.5°C to 15.5°C. One calorie is equal to 4.184 J. In the life sciences, heat is commonly measured in kilocalorie units, abbreviated Cal; 1 Cal is equal to 1000 cal.

9.3.2 Specific Heat

Specific heat is the quantity of heat required to raise the temperature of 1 g of a substance by 1 degree. The specific heats of some substances are shown in Table 9.1.

The human body is composed of water, proteins, fat, and minerals. Its specific heat reflects this composition. With 75% water and 25% protein, the specific heat of the body would be

$$\text{Specific heat} = 0.75 \times 1 + 0.25 \times 0.4 = 0.85$$

The specific heat of the average human body is closer to 0.83 due to its fat and mineral content, which we have not included in the calculation.

[2]For the symbol °C read degree Celsius. For the symbol C°, read Celsius degree.

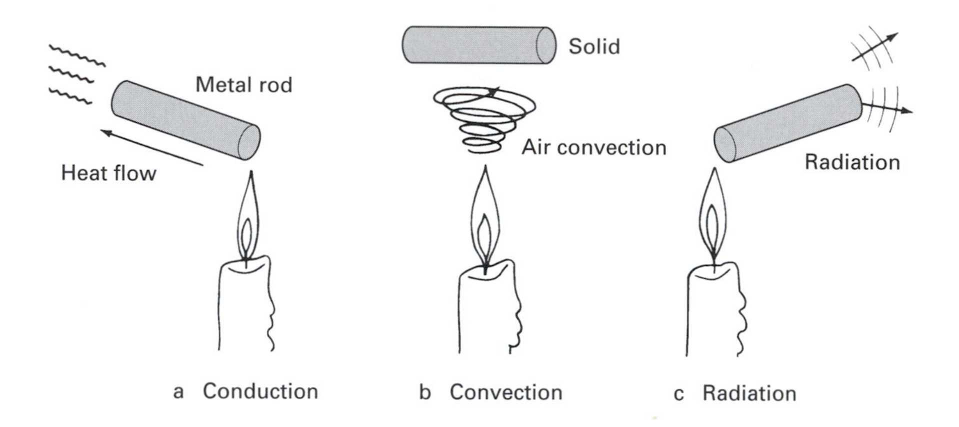

FIGURE 9.2 ▶ Heat is transferred from one region to another by (a) conduction, by (b) convection, and by (c) radiation.

9.3.3 Latent Heats

In order to convert a solid to a liquid at the same temperature or to convert a liquid to a gas, heat energy must be added to the substance. This energy is called *latent heat*. The latent heat of fusion is the amount of energy required to change 1 g of solid matter to liquid. The latent heat of vaporization is the amount of heat required to change 1 g of liquid to gas.

9.4 Transfer of Heat

Heat is transferred from one region to another in three ways: by conduction, convection, and radiation (Fig. 9.2).

9.4.1 Conduction

If one end of a solid rod is placed in the proximity of a heat source such as a fire, after some time the other end of the rod will become hot. In this case, heat has been transferred from the fire through the rod by conduction. The process of heat conduction involves the increase of internal energy in the material. The heat enters one end of the rod and increases the internal energy of the atoms near the heat source. In a solid material, the internal energy is in the vibration of the bound atoms and in the random motion of free electrons, which exist in some materials. The addition of heat increases both the random atomic vibrations and the speed of the electrons. The increased vibrational

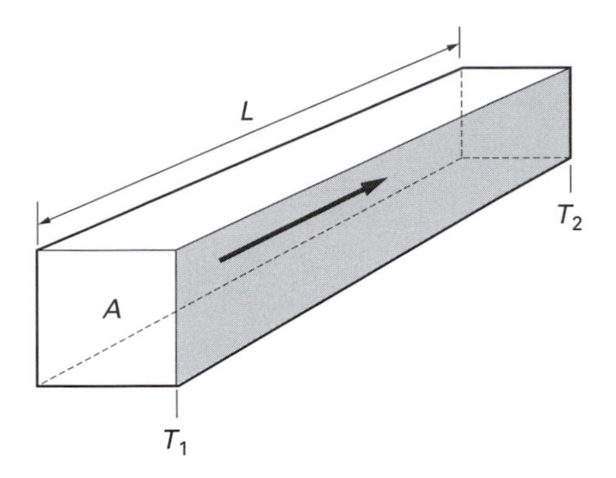

FIGURE 9.3 ▶ Heat flow through a block of material.

motion is transferred down the rod through collisions with neighboring atoms. However, because the atoms in the solid are tightly bound, their motion is restricted. Therefore, heat transfer via atomic vibrations is slow.

In some materials, the electrons in the atoms have enough energy to break loose from a specific nucleus and move freely through the material. The electrons move rapidly through the material so that, when they gain energy, they transfer it quickly to adjacent electrons and atoms. In this way, free electrons transfer the increase in the internal energy down the rod. Materials such as metals, which contain free electrons, are good conductors of heat; materials such as wood, which do not have free electrons, are insulators.

The amount of heat H_c conducted per second through a block of material (see Fig. 9.3) is given by

$$H_c = \frac{K_c A}{L}(T_1 - T_2) \tag{9.3}$$

Here A is the area of the block, L is its length, and $T_1 - T_2$ is the temperature difference between the two ends. The constant K_c is the *coefficient of thermal conductivity*. In physics texts, K_c is usually given in units of cal cm/sec-cm^2-C°. However, for problems involving living systems, it is often more convenient to express K_c in units of Cal cm/m^2-hr-C°. This is the amount of heat (in Cal units) per hour which flows through a slab of material 1 cm thick and 1 m square per C° temperature difference between the faces of the slab. The thermal conductivity of a few materials is given in Table 9.2.

9.4.2 Convection

In solids, heat transfer occurs by *conduction*; in fluids (gases and liquids), heat transfer proceeds primarily by *convection*. When a liquid or a gas is heated,

TABLE 9.2 ▶ Thermal Conductivity of Some Materials

Material	**Thermal conductivity, K_c** (Cal cm/m^2-hr-C°)
Silver	3.6×10^4
Cork	3.6
Tissue (unperfused)	18
Felt and down	0.36
Aluminum	1.76×10^4

the molecules near the heat source gain energy and tend to move away from the heat source. Therefore, the fluid near the heat source becomes less dense. Fluid from the denser region flows into the rarefied region, causing convection currents. These currents carry energy away from the heat source. When the energetic molecules in the heated convection current come in contact with a solid material, they transfer some of their energy to the atoms of the solid, increasing the internal energy of the solid. In this way, heat is coupled into a solid. The amount of heat transferred by convection per unit time H_c' is given by

$$H_c' = K_c' A(T_1 - T_2) \tag{9.4}$$

Here A is the area exposed to convective currents, $T_1 - T_2$ is the temperature difference between the surface and the convective fluid, and K_c' is the *coefficient of convection*, which is usually a function of the velocity of the convective fluid.

9.4.3 Radiation

Vibrating electrically charged particles emit electromagnetic radiation, which propagates away from the source at the speed of light. Electromagnetic radiation is itself energy (called *electromagnetic energy*), which in the case of a moving charge is obtained from the kinetic energy of the charged particle.

Because of internal energy, particles in a material are in constant random motion. Both the positively charged nuclei and the negatively charged electrons vibrate and, therefore, emit electromagnetic radiation. In this way, internal energy is converted into radiation, called *thermal radiation*. Due to the loss of internal energy, the material cools. The amount of radiation emitted by vibrating charged particles is proportional to the speed of vibration. Hot objects, therefore, emit more radiation than cold ones. Because the electrons are much lighter than the nuclei, they move faster and emit more radiant energy than the nuclei.

When a body is relatively cool, the radiation from it is in the long-wavelength region to which the eye does not respond. As the temperature (i.e., the *internal energy*) of the body increases, the wavelength of the radiation decreases. At high temperatures, some of the electromagnetic radiation is in the visible region, and the body is observed to glow.

When electromagnetic radiation impinges on an object, the charged particles (electrons) in the object are set into motion and gain kinetic energy. Electromagnetic radiation is, therefore, transformed into internal energy. The amount of radiation absorbed by a material depends on its composition. Some materials, such as carbon black, absorb most of the incident radiation. These materials are easily heated by radiation. Other materials, such as quartz and certain glasses, transmit the radiation without absorbing much of it. Metallic surfaces also reflect radiation without much absorption. Such reflecting and transmitting materials cannot be heated efficiently by radiation. The rate of emission of radiant energy H_r by a unit area of a body at temperature T is

$$H_r = e\sigma T^4 \tag{9.5}$$

Here σ is the *Stefan-Boltzmann constant*, which is 5.67×10^{-8} W/m^2-K^4 or 5.67×10^{-5} erg/cm^2-°K^4-sec. The temperature is measured on the absolute scale, and e is the *emissivity* of the surface, which depends on the temperature and nature of the surface. The value of the emissivity varies from 0 to 1. Emission and absorption of radiation are related phenomena; surfaces that are highly absorptive are also efficient emitters of radiation and have an emissivity close to 1. Conversely, surfaces that do not absorb radiation are poor emitters with a low value of emissivity.

A body at temperature T_1 in an environment at temperature T_2 will both emit and absorb radiation. The rate of energy emitted per unit area is $e\sigma T_1^4$, and the rate of energy absorbed per unit is $e\sigma T_2^4$. The values for e and σ are the same for both emission and absorption.

If a body at a temperature T_1 is placed in an environment at a lower temperature T_2, the net loss of energy from the body is

$$H_r = e\sigma \left(T_1^4 - T_2^4\right) \tag{9.6}$$

If the temperature of the body is lower than the temperature of the environment, the body gains energy at the same rate.

9.4.4 Diffusion

If a drop of colored solution is introduced into a still liquid, we observe that the color spreads gradually throughout the volume of the liquid. The

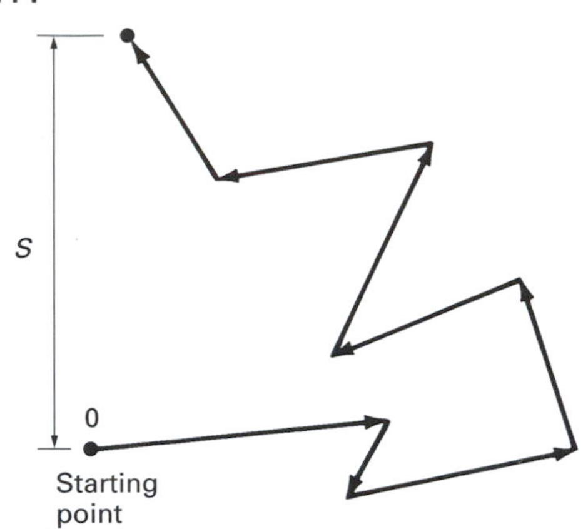

FIGURE 9.4 ▶ Random walk.

molecules of color spread from the region of high concentration (of the initially introduced drop) to regions of lower concentration. This process is called *diffusion*.

Diffusion is the main mechanism for the delivery of oxygen and nutrients into cells and for the elimination of waste products from cells. On a large scale, diffusive motion is relatively slow (it may take hours for the colored solution in our example to diffuse over a distance of a few centimeters), but on the small scale of tissue cells, diffusive motion is fast enough to provide for the life function of cells.

Diffusion is the direct consequence of the random thermal motion of molecules. Although a detailed treatment of diffusion is beyond our scope, some of the features of diffusive motion can be deduced from simple kinetic theory.

Consider a molecule in a liquid or a gas which is moving away from the starting point 0. The molecule has a thermal velocity v and travels on the average a distance L before colliding with another molecule (see Fig. 9.4). As a result of the collision, the direction of motion of the molecule is changed randomly. The path may be only slightly deflected, or it may be altered substantially. On the average, however, after a certain number of collisions the molecule will be found a distance S from the starting point. A statistical analysis of this type of motion shows that after N collisions the distance of the molecule from the starting point is, on the average,

$$S = L\sqrt{N} \tag{9.7}$$

The average distance (L) traveled between collisions is called the *mean free path*. This type of diffusive motion is called a *random walk*.

A frequently used illustration of the random walk examines the position of a drunkard walking away from a lamppost. He starts off in a particular direction, but with each step he changes his direction of motion randomly.

If the length of each step is 1 m, after taking 100 steps he will be only 10 m away from the lamppost although he has walked a total of 100 m. After 10,000 steps, having walked 10 km, he will be still only 100 m (on the average) from his starting point.

Let us now calculate the length of time required for a molecule to diffuse a distance S from the starting point. From Eq. 9.7 the number of steps or collisions that take place while diffusing through a distance S is

$$N = \frac{S^2}{L^2} \tag{9.8}$$

The total distance traveled is the product of the number of steps and the length of each step; that is,

$$\text{Total distance} = NL = \frac{S^2}{L} \tag{9.9}$$

If the average velocity of the particle is v, the time t required to diffuse a distance S is

$$t = \frac{\text{Total distance}}{v} = \frac{S^2}{Lv} \tag{9.10}$$

Although our treatment of diffusion has been simplified, Eq. 9.10 does lead to reasonable estimates of diffusion times. In a liquid such as water, molecules are close together. Therefore, the mean free path of a diffusing molecule is short, about 10^{-8} cm (this is approximately the distance between atoms in a liquid). The velocity of the molecule depends on the temperature and on its mass. At room temperature, the velocity of a light molecule may be about 10^4 cm/sec. From Eq. 9.10, the time required for molecules to diffuse a distance of 1 cm is

$$t = \frac{S^2}{Lv} = \frac{(1)^2}{10^{-8} \times 10^4} = 10^4 \text{ sec} = 2.8 \text{ hr}$$

However, the time required to diffuse a distance of 10^{-3} cm, which is the typical size of a tissue cell, is only 10^{-2} sec (see Exercise 9-3a).

Gases are less densely packed than liquids; consequently, in gases the mean free path is longer and the diffusion time shorter. In a gas at 1 atm pressure, the mean free path is on the order of 10^{-5}—the exact value depends on the specific gas. The time required to diffuse a distance of 1 cm is about 10 sec. Diffusion through a distance of 10^{-3} cm takes only 10^{-5} sec (see Exercise 9-3b).

9.5 Transport of Molecules by Diffusion

We will now calculate the number of molecules transported by diffusion from one region to another. Consider a cylinder containing a nonuniform distribution of diffusing molecules or other small particles (see Fig. 9.5). At position $x = 0$, the density of the diffusing molecules is C_1. At a small distance Δx away from this point, the concentration is C_2. We can define a diffusion velocity V_D as the average speed of diffusion from $x = 0$ to $x = \Delta x$. This velocity is simply the distance Δx divided by the average time for diffusion t; that is,

$$V_D = \frac{\Delta x}{t}$$

Substituting $t = (\Delta x)^2/Lv$ from Eq. 9.10, we obtain

$$V_D = \frac{\Delta x}{(\Delta x)^2/Lv} = \frac{Lv}{\Delta x} \tag{9.11}$$

(Remember that v here is the thermal velocity.) The number of molecules J arriving per second per unit area, from region 1 where the density is C_1 to region 2 (see Exercise 9-4) is

$$J_1 = \frac{V_D C_1}{2} \tag{9.12}$$

The factor of 2 in the denominator accounts for the fact that molecules are diffusing both toward and away from region 2. The term J is called the *flux*, and it is in units ($\text{cm}^{-2}\text{s}^{-1}$).

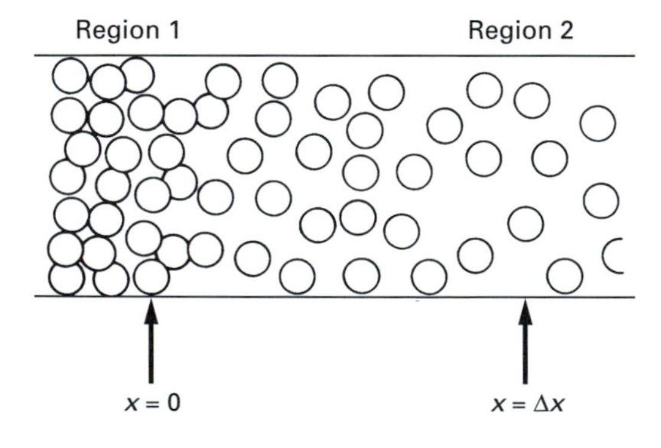

FIGURE 9.5 ▶ Diffusion.

At the same time, molecules are also diffusing from region 2 where the density is C_2 toward region 1. This flux J_2 is

$$J_2 = \frac{V_D C_2}{2}$$

The net flux of molecules into region 2 is the difference between the arriving and the departing flux, which is

$$J = J_1 - J_2 = \frac{V_D (C_1 - C_2)}{2}$$

Substituting for $V_D = Lv/\Delta x$, we obtain

$$J = \frac{Lv (C_1 - C_2)}{2\Delta x} \tag{9.13}$$

This derivation assumes that the velocities v in the two regions are the same. Although this solution for the diffusion problem is not exact, it does illustrate the nature of the diffusion process. (For a more rigorous treatment, see, for example, [11-7]). The net flux from one region to another depends on the difference in the density of the diffusing particles in the two regions. The flux increases with thermal velocity v and decreases with the distance between the two regions.

Equation 9.13 is usually written as

$$J = \frac{D}{\Delta x} (C_1 - C_2) \tag{9.14}$$

where D is called the *diffusion coefficient*. In our case, the diffusion coefficient is simply

$$D = \frac{Lv}{2} \tag{9.15}$$

In general, however, the diffusion coefficient is a more complex function because the mean free path L depends on the size of the molecule and the viscosity of the diffusing medium. In our previous illustration of diffusion through a fluid, where $L = 10^{-8}$ cm and $v = 10^4$ cm/sec, the diffusion coefficient calculated from Eq. 9.15 is 5×10^{-5} cm^2/sec. By comparison, the measured diffusion coefficient of salt (NaCl) in water, for example, is 1.09×10^{-5} cm^2/sec. Thus, our simple calculation gives a reasonable estimate for the diffusion coefficient. Larger molecules, of course, have a smaller diffusion coefficient. The diffusion coefficients for biologically important molecules are in the range from 10^{-7} to 10^{-6} cm^2/sec.

9.6 Diffusion through Membranes

So far we have discussed only free diffusion through a fluid, but the cells constituting living systems are surrounded by membranes which impede free diffusion. Oxygen, nutrients, and waste products must pass through these membranes to maintain the life functions. In the simplest model, the biological membrane can be regarded as porous, with the size and the density of the pores governing the diffusion through the membrane. If the diffusing molecule is smaller than the size of the pores, the only effect of the membrane is to reduce the effective diffusion area and thus decrease the diffusion rate. If the diffusing molecule is larger than the size of the pores, the flow of molecules through the membrane may be barred. (See Fig. 9.6.) (Some molecules may still get through the membrane, however, by dissolving into the membrane material.)

The net flux of molecules J flowing through a membrane is given in terms of the permeability of the membrane P

$$J = P(C_1 - C_2) \tag{9.16}$$

This equation is similar to Eq. 9.14 except that the term D is replaced by the permeability P, which includes the diffusion coefficient as well as the effective thickness Δx of the membrane. The permeability depends, of course, on the type of membrane as well as on the diffusing molecule. Permeability may be nearly zero (if the molecules cannot pass through the membrane) or as high as 10^{-4} cm/sec.

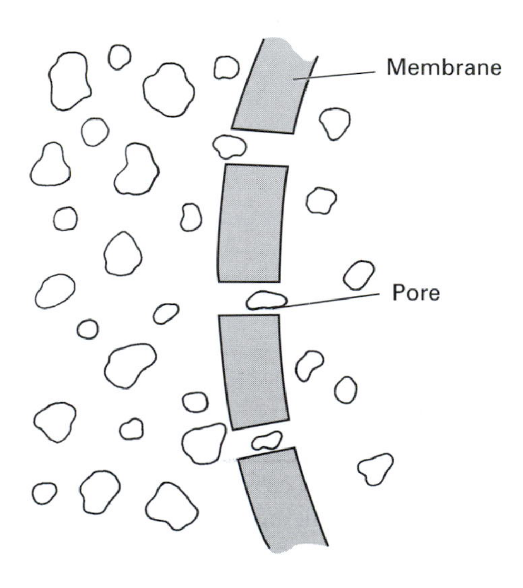

FIGURE 9.6 ▶ Diffusion through a membrane.

The dependence of permeability on the diffusing species allows the cell to maintain a composition different from that of the surrounding environment. Many membranes, for example, are permeable to water but do not pass molecules dissolved in water. As a result water can enter the cell, but the components of the cell cannot pass out of the cell. Such a one-way passage of water is called *osmosis*.

In the type of diffusive motion we have discussed so far, the movement of the molecules is due to their thermal kinetic energy. Some materials, however, are transported through membranes with the aid of electric fields that are generated by charge differences across the membrane. This type of a transport will be discussed in Chapter 13.

We have shown that over distances larger than a few millimeters diffusion is a slow process. Therefore, large living organisms must use circulating systems to transport oxygen nutrients and waste products to and from the cells. The evolution of the respiratory system in animals is a direct consequence of the inadequacy of diffusive transportation over long distances.

9.7 The Respiratory System

As will be shown in the following two chapters, animals require energy to function. This energy is provided by food, which is oxidized by the body. On the average, 0.207 liter of oxygen at 760 torr are required for every Cal of energy released by the oxidation of food in the body. At rest, an average 70-kg adult requires about 70 Cal of energy per hour, which implies a consumption of 14.5 liter of O_2 per hour, which is about 10^{20} oxygen molecules per second (see Exercise 9-5).

The simplest way to obtain the required oxygen is by diffusion through the skin. This method, however, cannot supply the needs of large animals. It has been determined that in a person only about 2% of oxygen consumed at rest is obtained by diffusion through the skin. The rest of the oxygen is obtained through the lungs.

The lungs can be thought of as an elastic bag suspended in the chest cavity (see Fig. 9.7). When the diaphragm descends, the volume of the lungs increases, causing a reduction in gas pressure inside the lungs. As a result, air enters the lungs through the trachea. The trachea branches into smaller and smaller tubes, which finally terminate at tiny cavities called *alveoli*. It is here that gas is exchanged by diffusion between the blood and the air in the lungs. The lungs of an adult contain about 300 million alveoli with diameters ranging between 0.1 and 0.3 mm. The total alveolar area of the lungs is about 100 m^2, which is about 50 times larger than the total surface area of the skin. The barrier between the alveolar air and the blood in the capillaries is very

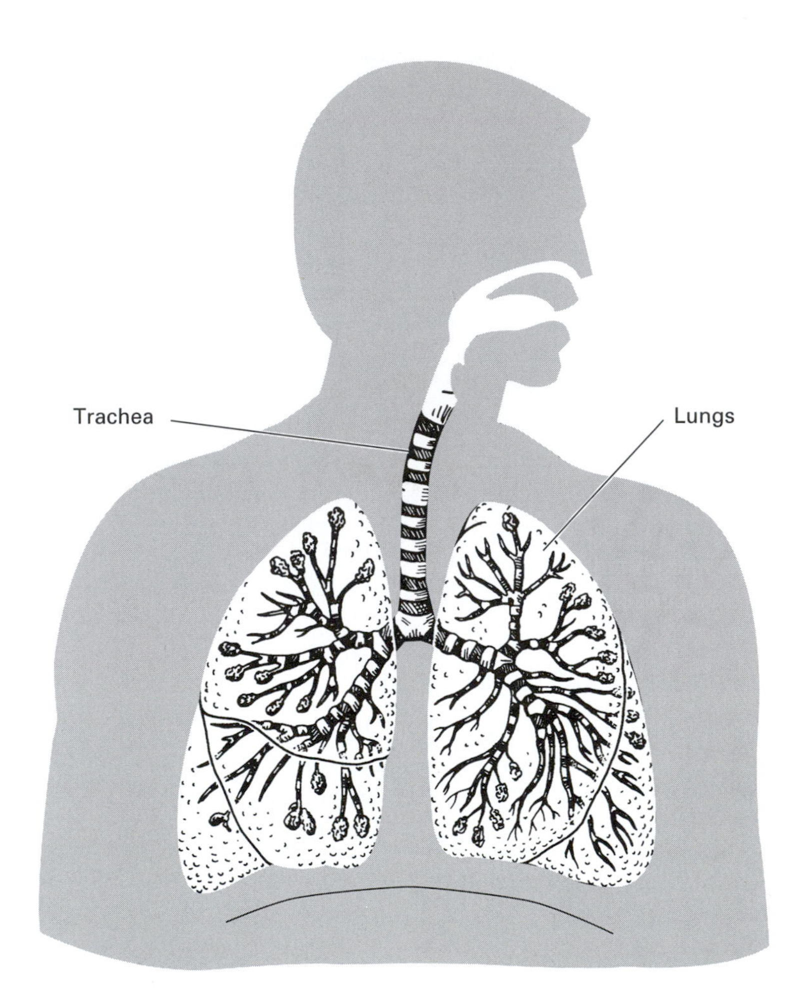

FIGURE 9.7 ▶ Lungs.

thin, only about 4×10^{-5} cm. Therefore, the gas exchange of oxygen into the blood and CO_2 out of the blood is very fast.

The lungs are not fully emptied and filled with each breath. In fact, the full volume of the lungs is about 6 liter, and at rest only about $\frac{1}{2}$ liter is exchanged during each breath. The composition of inspired and expired air is shown in Table 9.3.

Using the experimental data in Table 9.3, we can easily show that about 10.5 breaths per minute satisfy the oxygen requirements for a resting person (see Exercise 9-6). The oxygen requirement, of course, rises with increased

TABLE 9.3 ▶ The Percentage of N_2, O_2, and CO_2 in Inspired and Expired Air for a Resting Person

	N_2	O_2	CO_2
Inspired air	79.02	20.94	0.04
Expired air	79.2	16.3	4.5

physical activity, which results in both faster and deeper breathing. During deep breathing, as much as 70% of the air in the lungs is exchanged in each breath.

While diffusion through the skin can supply only a small fraction of the oxygen required by large animals, the oxygen needs of small animals may be completely satisfied through this channel. This can be deduced from the following considerations. The energy consumption and, hence, the oxygen requirement of an animal is approximately proportional to its mass.[3] The mass in turn is proportional to the volume of the animal. The amount of oxygen diffusing through the skin is proportional to the surface area of the skin. Now, if R is a characteristic linear dimension of the animal, the volume is proportional to R^3, and the skin surface area is proportional to R^2. The surface to volume ratio is given by

$$\frac{\text{Surface area}}{\text{Volume}} = \frac{R^2}{R^3} = \frac{1}{R} \tag{9.17}$$

Therefore, as the size of the animal R decreases, its surface-to-volume ratio increases; that is, for a unit volume, a small animal has a greater surface area than a large animal.

It is possible to obtain an estimate for the maximum size of the animal that can get its oxygen entirely by skin diffusion. A highly simplified calculation outlined in Exercise 9-7 shows that the maximum linear size of such an animal is about 0.5 cm. Therefore, only small animals, such as insects, can rely entirely on the diffusion transfer to provide them with oxygen. However, during hibernation when the oxygen requirements of the animal are reduced to a very low value, larger animals such as frogs can obtain all the necessary oxygen through their skin. In fact some species of frog hibernate through the winter at the bottom of lakes where the temperature is constant at 4°C. The required oxygen enters the frog's body by diffusion from the surrounding water, which contains dissolved oxygen.

---

[3] This is an approximation. A more detailed discussion is found in [11-10].

9.8 Diffusion and Contact Lenses

Most parts of the human body receive the required oxygen from the circulating blood. However, the *cornea*, which is the transparent surface layer of the eye, does not contain blood vessels (this allows it to be transparent). The cells in the cornea receive oxygen by diffusion from the surface layer of tear fluid, which contains oxygen. This fact allows us to understand why most contact lenses should not be worn during sleep. The contact lens is fitted so that blinking rocks the lens slightly. This rocking motion brings fresh oxygen-rich tear fluid under the lens. Of course, when people sleep they do not blink; therefore, the corneas under their contact lenses are deprived of oxygen. This may result in a loss of corneal transparency.

▶ EXERCISES ▶

9-1. Fish using air bladders to control their buoyancy are less stable than those using porous bones. Explain this phenomenon using the gas equation (Eq. 9.2). (*Hint*: What happens to the air bladder as the fish sinks to a greater depth?)

9-2. A scuba diver breathes air from a tank which has a pressure regulator that automatically adjusts the pressure of the inhaled air to the ambient pressure. If a diver 40 m below the surface fills his lungs to the full capacity of 6 liters and then rises quickly to the surface, to what volume will his lungs expand? Is such a rapid ascent advisable?

9-3. (a) Calculate the time required for molecules to diffuse in a liquid a distance of 10^{-3} cm. Assume that the average velocity of the molecules is 10^4 cm/sec and that the mean free path is 10^{-8} cm. (b) Repeat the calculation for diffusion in a gas where the mean free path is 10^{-5} cm.

9-4. Consider a beam of particles traveling at a velocity V_D. If the area of the beam is A and the density of particle in the beam is C, show that the number of particles that pass by a given point each second is $V_D \times C \times A$.

9-5. A consumption of 14.5 liters of oxygen per hour is equivalent to how many molecules per second? (The number of molecules per cubic centimeter at 0°C and 760 torr is 2.69×10^{19}.)

9-6. Using the data in the text and in Table 9.3, calculate the number of breaths per minute required to satisfy the oxygen needs of a resting person.

9-7. (a) We stated in the text that the oxygen consumption at rest for a 70-kg person is 14.5 liter/hr and that 2% of this requirement is provided by the diffusion of oxygen through the skin. Assuming that the skin surface area of the person is 1.7 m^2, calculate the diffusion rate for oxygen through the skin in liter/hr-cm^2. (b) What is the maximum linear size of an animal whose oxygen requirements at rest can be provided by diffusion through the skin?

Use the following assumptions:

(i) The density of animal tissue is 1 g/cm^3.

(ii) Per unit volume, all animals require the same amount of oxygen.

(iii) The animal is spherical in shape.

Chapter 10

Thermodynamics

Thermodynamics is the study of the relationship between heat, work, and the associated flow of energy. After many decades of experience with heat phenomena, scientists formulated two fundamental laws as the foundation of thermodynamics. The First Law of Thermodynamics states that energy, which includes heat, is conserved; that is, one form of energy can be converted into another, but energy can neither be created nor destroyed. This implies that the total amount of energy in the universe is a constant.[1] The second law, more complex than the first, can be stated in a number of ways which, although they appear different, can be shown to be equivalent. Perhaps the simplest statement of the Second Law of Thermodynamics is that *spontaneous change in nature occurs from a state of order to a state of disorder*.

10.1 First Law of Thermodynamics

One of the first to state the law of energy conservation was the German physician Robert Mayer (1814–1878). In 1840 Mayer was the physician on the schooner *Java*, which sailed for the East Indies. While aboard ship, he was reading a treatise by the French scientist Laurent Lavoisier in which Lavoisier suggested that the heat produced by animals is due to the slow combustion of food in their bodies. Lavoisier further noted that less food is burned by the body in a hot environment than in a cold one.

[1] It has been shown by the theory of relativity that the conservation law must include matter which is convertible to energy.

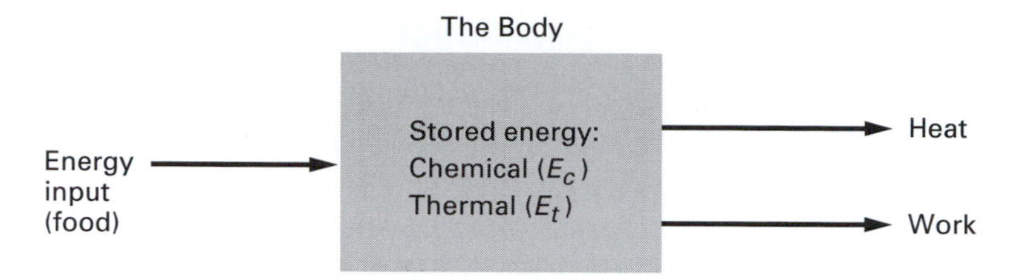

FIGURE 10.1 ▶ The energetics of the body.

When the ship reached the tropics, many of its crew became sick with fever. Applying the usual remedy for fever, Mayer bled his patients. He noticed that the venous blood, which is normally dark red, was nearly as red as arterial blood. He considered this a verification of Lavoisier's suggestion. Because in the tropics less fuel is burned in the body, the oxygen content of the venal blood is high, giving it the brighter color. Mayer then went beyond Lavoisier's theory and suggested that in the body there is an exact balance of energy (which he called *force*). The energy released by the food is balanced by the lost body heat and the work done by the body. Mayer wrote in an article published in 1842, "Once in existence, force [energy] cannot be annihilated—it can only change its form."

Considerably more evidence had to be presented before conservation of energy was accepted as a law, but it is interesting that such a fundamental physical law was first suggested from the observation of human physiology.

Conservation of energy is implicit in all our calculations of energy balance in living systems. Consider, for example, the energetics for the functioning of an animal (see Fig. 10.1). The body of an animal contains internal thermal energy E_t, which is the product of the mass and specific heat, and chemical energy E_c stored in the tissue of the body. In terms of energy, the activities of an animal consist of simply eating, working, and rejecting excess heat by means of various cooling mechanisms (radiation, convection, etc.). Without going into detailed calculations, the first law allows us to draw some conclusions about the energetics of the animal. For example, if the internal temperature and the weight of the animal are to remain constant (i.e., E_c and E_t constant), over a given period of time the energy intake must be exactly equal to the sum of the work done and the heat lost by the body. An imbalance between intake and output energy implies a change in the sum $E_c + E_t$. The First Law of Thermodynamics is implicit in all the numerical calculations presented in Chapter 11.

10.2 Second Law of Thermodynamics

There are many imaginable phenomena that are not forbidden by the First Law of Thermodynamics but still do not occur. For example, when an object falls from a table to the ground, its potential energy is first converted into kinetic energy; then, as the object comes to rest on the ground, the kinetic energy is converted into heat. The First Law of Thermodynamics does not forbid the reverse process, whereby the heat from the floor would enter the object and be converted into kinetic energy, causing the object to jump back on the table. Yet this event does not occur. Experience has shown that certain types of events are irreversible. Broken objects do not mend by themselves. Spilled water does not collect itself back into a container. The irreversibility of these types of events is intimately connected with the probabilistic behavior of systems comprised of a large ensemble of subunits.

As an example, consider three coins arranged heads up on a tray. We will consider this an ordered arrangement. Suppose that we now shake the tray so that each coin has an equal chance of landing on the tray with either head or tail up. The possible arrangements of coins that we may obtain are shown in Table 10.1. Note that there are eight possible outcomes of tossing the three coins. Of these, only one yields the original ordered arrangement of three heads (H,H,H). Because the probabilities of obtaining any one of the coin arrangements in Table 10.1 are the same, the probability of obtaining the three-head arrangement after shaking the tray once is $1/8$, or 0.125; that is, on the average, we must toss the coins eight times before we can expect to see the three-head arrangement again.

As the number of coins in the experiment is increased, the probability of returning to the ordered arrangement of all heads decreases. With 10 coins on the tray, the probability of obtaining all heads after shaking the tray is 0.001. With 1000 coins, the probability of obtaining all heads is so small as

TABLE 10.1 ► The Ordering of Three Coins

Coin 1	Coin 2	Coin 3
H	H	H
H	H	T
H	T	H
T	H	H
H	T	T
T	H	T
T	T	H
T	T	T

to be negligible. We could shake the tray for many years without seeing the ordered arrangement again. In summary, the following is to be noted from this illustration: The number of possible coin arrangements is large, and only one of them is the ordered arrangement; therefore, although any one of the coin arrangements—including the ordered one—is equally likely, the probability of returning to an ordered arrangement is small. As the number of coins in the ensemble increases, the probability of returning to an ordered arrangement decreases. In other words, if we disturb an ordered arrangement, it is likely to become disordered. This type of behavior is characteristic of all events that involve a collective behavior of many components.

The Second Law of Thermodynamics is a statement about the type of probabilistic behavior illustrated by our coin experiment. One statement of the second law is: *The direction of spontaneous change in a system is from an arrangement of lesser probability to an arrangement of greater probability*; that is, from order to disorder. This statement may seem to be so obvious as to be trivial, but, once the universal applicability of the second law is recognized, its implications are seen to be enormous. We can deduce from the second law the limitations on information transmission, the meaning of time sequence, and even the fate of the universe. These subjects, however, are beyond the scope of our discussion.

One important implication of the second law is the limitation on the conversion of heat and internal energy to work. This restriction can be understood by examining the difference between heat and other forms of energy.

10.3 Difference between Heat and Other Forms of Energy

We defined heat as energy being transferred from a hotter to a colder body. Yet when we examined the details of this energy transfer, we saw that it could be attributed to transfer of a specific type of energy such as kinetic, vibrational, electromagnetic, or any combination of these (see Chapter 9). For this reason, it may not seem obvious why the concept of heat is necessary. It is, in fact, possible to develop a theory of thermodynamics without using the concept of heat explicitly, but we would then have to deal with each type of energy transfer separately, and this would be difficult and cumbersome. In many cases, energy is being transferred to or from a body by different methods, and keeping track of each of these is often not possible and usually not necessary. No matter how energy enters the body, its effect is the same. It raises the internal energy of the body. The concept of heat energy is, therefore, very useful.

The main feature that distinguishes heat from other forms of energy is the random nature of its manifestations. For example, when heat flows via con-

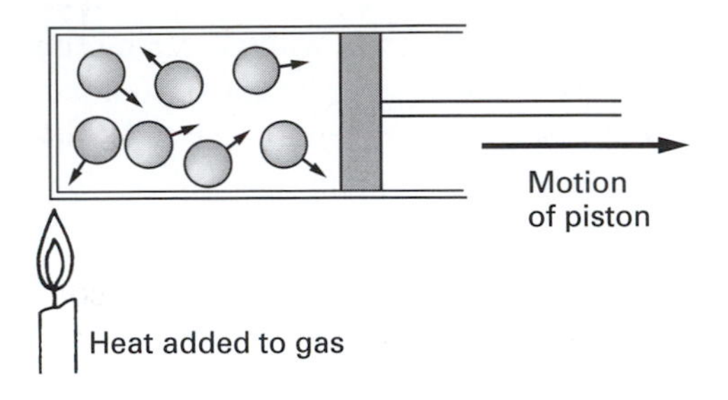

FIGURE 10.2 ▶ The motion of a piston.

duction from one part of the material to another part, the flow occurs through the sequential increase in the internal energy along the material. This internal energy is in the form of random chaotic motion of atoms. Similarly, when heat is transferred by radiation, the propagating waves travel in random directions. The radiation is emitted over a wide wavelength (color) range, and the phases of the wave along the wave front are random. By comparison, other forms of energy are more ordered. Chemical energy, for example, is present by virtue of specific arrangements of atoms in a molecule. Potential energy is due to the well-defined position, or configuration, of an object.

While one form of energy can be converted to another, heat energy, because of its random nature, cannot be completely converted to other forms of energy. We will use the behavior of a gas to illustrate our discussion. First, let us examine how heat is converted to work in a heat engine (for example, the steam engine). Consider a gas in a cylinder with a piston (see Fig. 10.2). Heat flows into the gas; this increases the kinetic energy of the gas molecules and, therefore, raises the internal energy of the gas. The molecules moving in the direction of the piston collide with the piston and exert a force on it. Under the influence of this force, the piston moves. In this way, heat is converted into work via internal energy.

The heat added to the gas causes the molecules in the cylinder to move in random directions, but only the molecules that move in the direction of the piston can exert a force on it. Therefore, the kinetic energy of only the molecules that move toward the piston can be converted into work. For the added heat to be completely converted into work, all the gas molecules would have to move in the direction of the piston motion. In a large ensemble of molecules, this is very unlikely.

The odds against the complete conversion of 1 cal of heat into work can be expressed in terms of a group of monkeys who are hitting typewriter keys at random and who by chance type out the complete works of Shakespeare without error. The probability that 1 cal of heat would be completely converted to work is about the same as the probability that the monkeys would

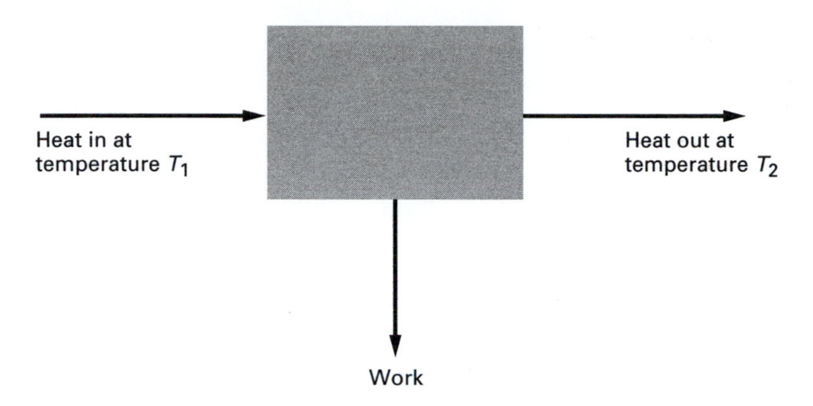

FIGURE 10.3 ▶ Conversion of heat to work.

type Shakespeare's works 15 quadrillion times in succession. (This example is taken from [11-2].)

The distinction between work and heat is this: In work, the energy is in an ordered motion; in heat, the energy is in random motion. Although some of the random thermal motion can be ordered again, the ordering of all the motion is very improbable. Because the probability of completely converting heat to work is vanishingly small, the Second Law of Thermodynamics states categorically that it is impossible.

Heat can be partially converted to work as it flows from a region of higher temperature T_1 to a region of lower temperature T_2 (see Fig. 10.3). A quantitative treatment of thermodynamics shows (see, for example, [11-5]) that the maximum ratio of work to the input heat is

$$\frac{\text{Work}}{\text{Heat input}} = 1 - \frac{T_2}{T_1} \tag{10.1}$$

Here the temperature is measured on the absolute scale.

From this equation, it is evident that heat can be completely converted into work only if the heat is rejected into a reservoir at absolute zero temperature. Although objects can be cooled to within a very small fraction of absolute zero, absolute zero cannot be attained. Therefore, heat cannot be completely converted into work.

10.4 Thermodynamics of Living Systems

It is obvious that animals need food to live, but the reason for this is less obvious. The idea that animals need energy because they consume energy is, strictly speaking, incorrect. We know from the First Law of Thermodynamics that energy is conserved. The body does not consume energy, it changes it from one form to another. In fact, the first law could lead us to the erroneous conclusion that animals should be able to function without a source of external energy. The body takes in energy that is in the chemical bonds of the food molecules and converts it to heat. If the weight and the temperature of the body remain constant and if the body performs no external work, the energy input to the body equals exactly the heat energy leaving the body. We may suppose that if the heat outflow could be stopped—by good insulation, for example—the body could survive without food. As we know, this supposition is wrong. The need for energy is made apparent by examining the functioning of the body in the light of the Second Law of Thermodynamics.

The body is a highly ordered system. A single protein molecule in the body may consist of a million atoms bound together in an ordered sequence. Cells are more complex still. Their specialized functions within the body depend on a specific structure and location. We know from the Second Law of Thermodynamics that such a highly ordered system, left to itself, tends to become disordered, and once it is disordered, it ceases to function. Work must be done on the system continuously to prevent it from falling apart. For example, the blood circulating in veins and arteries is subject to friction, which changes kinetic energy to heat and slows the flow of blood. If a force were not applied to the blood, it would stop flowing in a few seconds. The concentration of minerals inside a cell differs from that in the surrounding environment. This represents an ordered arrangement. The natural tendency is toward an equalization with the environment. Work must be done to prevent the contents of the cell from leaking out. Finally, cells that die must be replaced, and if the animal is growing, new tissue must be manufactured. For such replacement and growth, new proteins and other cell constituents must be put together from smaller, relatively more random subcomponents. Thus, the process of life consists of building and maintaining ordered structures. In the face of the natural tendency toward disorder, this activity requires work. The situation is somewhat analogous to a pillar made of small, slippery, uneven blocks that tend to slide out of the structure. The pillar remains standing only if blocks are continuously pushed back.

The work necessary to maintain the ordered structures in the body is obtained from the chemical energy in food. Except for the energy utilized in external work done by the muscles, all the energy provided by food is ultimately converted into heat by friction and other dissipative processes in the body. Once the temperature of the body is at the desired level, all the heat generated by the body must leave through the various cooling mechanisms of the body (see Chapter 11). The heat must be dissipated because, unlike heat engines (such as the turbine or the steam engine), the body does not have the ability to obtain work from heat energy. The body can obtain work only from chemical energy. Even if the body did have mechanisms for using heat to perform work, the amount of work it could obtain in this way would be small. Once again, the second law sets the limit. The temperature differences in the body are small—not more than about 7 C° between the interior and the exterior. With the interior temperature T_1 at 310 K (37 °C) and the exterior temperature T_1 at 303 K, the efficiency of heat conversion to work would be (from Eq. 10.1) at most only about 2%.

Of all the various forms of energy, the body can utilize only the chemical binding energy of the molecules which constitute food. The body does not have a mechanism to convert the other forms of energy into work. A person could bask in the sun indefinitely, receiving large quantities of radiant energy, and yet die of starvation. Plants, on the other hand, are able to utilize radiant energy. As animals use chemical energy, so plants utilize solar radiation to provide the energy for the ordering processes necessary for life.

The organic materials produced in the life cycle of plants provide food energy for herbivorous animals, which in turn are food for the carnivorous animals that eat them. The sun is, thus, the ultimate source of energy for life on Earth.

Since living systems create order out of relative disorder (for example, by synthesizing large complex molecules out of randomly arranged subunits), it may appear at first glance that they violate the Second Law of Thermodynamics, but this is not the case. To ascertain that the second law is valid, we must examine the whole process of life, which includes not only the living unit but also the energy that it consumes and the by-products that it rejects. To begin with, the food that is consumed by an animal contains a considerable degree of order. The atoms in the food molecules are not randomly arranged but are ordered in specific patterns. When the chemical energy in the molecular bindings of the food is released, the ordered structures are broken down. The eliminated waste products are considerably more disordered than the food taken in. The ordered chemical energy is converted by the body into disordered heat energy.

The amount of disorder in a system can be expressed quantitatively by means of a concept called *entropy*. Calculations show that, in all cases, the increase in the entropy (disorder) in the surroundings produced by the living system is always greater than the decrease in entropy (i.e., ordering) obtained in the living system itself. The total process of life, therefore, obeys the second law. Thus, living systems are perturbations in the flow toward disorder. They keep themselves ordered for a while at the expense of the environment. This is a difficult task requiring the use of the most complex mechanisms found in nature. When these mechanisms fail, as they eventually must, the order falls apart, and the organism dies.

10.5 Information and the Second Law

We have stressed earlier that work must be done to create and maintain the highly ordered local state of life. We now turn to the question, what else is needed for such local ordering to occur? Perhaps we can get an insight into this issue from a simple everyday experience. In the course of time, our apartment becomes disordered. Books, which had been placed neatly, in alphabetical order, on a shelf in the living room, are now strewn on the table and some are even under the bed. Dishes that were clean and neatly stacked in the cupboard, are now dirty with half-eaten food and are on the living room table. We decide to clean up, and in 15 minutes or so the apartment is back in order. The books are neatly shelved, and the dishes are clean and stacked in the kitchen. The apartment is clean.

Two factors were necessary for this process to occur. First, as was already stated, energy was required to do the work of gathering and stacking the books and cleaning and ordering the dishes. Second, and just as important, information was required to direct the work in the appropriate direction. We had to know where to place the books and how to clean the dishes and stack them just so. The concept of information is of central importance here.

In the 1940s, Claude Shannon developed a quantitative formulation for the amount of information available in a given system. Shannon's formula for information content is shown to be equivalent to the formula for entropy—the measure of disorder—except, with a negative sign. This mathematical insight formally shows that if energy and information are available, the entropy in a given locality can be decreased by the amount of information available to engage in the process of ordering. In other words, as in our example of the messy living room, order can be created in a disordered system by work that is directed by appropriate information. The second law, of course, remains valid: *the overall entropy of the universe increases*. The work required to perform the ordering, one way or another, causes a greater disorder in the surroundings than the order that was created in the system itself. It is the availability of information and energy that allows living systems to replicate, grow, and maintain their structures.

The chain of life begins with plants that possess information in their genetic material on how to utilize the energy from the sun to make highly ordered complex structures from the simple molecules available to them: principally water, carbon dioxide, and an assortment of minerals. The process is, in essence, similar in human beings and other animals. All the information required for the function of the organism is contained in the intricate structure of DNA. The human DNA consists of about a billion molecular units in a well-determined sequence. Utilizing the energy obtained from the food that is consumed by the organism, the information in the DNA guides the assembly of the various proteins and enzymes required for the functioning of the organism.

▶ EXERCISES ▶

10-1. Explain how the second law of thermodynamics limits conversion of heat to work.

10-2. From your own experience, give an example of the second law of thermodynamics.

10-3. Describe the connections between information, the second law of thermodynamics, and living systems.

Chapter 11

Heat and Life

The degree of hotness, or temperature, is one of the most important environmental factors in the functioning of living organisms. The rates of the metabolic processes necessary for life, such as cell divisions and enzyme reactions, depend on temperature. Generally the rates increase with temperature. A 10 degree change in temperature may alter the rate by a factor of 2.

Because liquid water is an essential component of living organisms as we know them, the metabolic processes function only within a relatively narrow range of temperatures, from about 2°C to 120°C. Only the simplest of living organisms can function near the extremes of this range.[1] Large-scale living systems are restricted to a much narrower range of temperatures.

The functioning of most living systems, plants and animals, is severely limited by seasonal variations in temperature. The life processes in reptiles, for example, slow down in cold weather to a point where they essentially cease to function. On hot sunny days these animals must find shaded shelter to keep their body temperatures down.

For a given animal, there is usually an optimum rate for the various metabolic processes. Warm-blooded animals (mammals and birds) have evolved methods for maintaining their internal body temperatures at near constant levels. As a result, warm-blooded animals are able to function at an optimum level over a wide range of external temperatures. Although this temperature regulation requires additional expenditures of energy, the adaptability achieved is well worth this expenditure.

[1] In deep oceans, the pressure is high and so is the boiling point of water. Here certain *thermophilic* bacteria can survive near thermal vents at significantly higher temperatures.

TABLE 11.1 ▶ Metabolic Rates for Selected Activities

Activity	Metabolic rate (Cal/m^2-hr)
Sleeping	35
Lying awake	40
Sitting upright	50
Standing	60
Walking (3 mph)	140
Moderate physical work	150
Bicycling	250
Running	600
Shivering	250

In this chapter, we will examine energy consumption, heat flow, and temperature control in animals. Although most of our examples will be specific to people, the principles are generally applicable to all animals.

11.1 Energy Requirements of People

All living systems need energy to function. In animals, this energy is used to circulate blood, obtain oxygen, repair cells, and so on. As a result, even at complete rest in a comfortable environment, the body requires energy to sustain its life functions. For example, a man weighing 70 kg lying quietly awake consumes about 70 Cal/hr (1 Cal/hr = 1.16 W). Of course, the energy expenditure increases with activity.

The amount of energy consumed by a person depends on the person's weight and build. It has been found, however, that the amount of energy consumed by a person during a given activity divided by the surface area of the person's body is approximately the same for most people. Therefore, the energy consumed for various activities is usually quoted in Cal/m^2-hr. This rate is known as the *metabolic rate*. The metabolic rates for some human activities are shown in Table 11.1. To obtain the total energy consumption per hour, we multiply the metabolic rate by the surface area of the person. The following empirical formula yields a good estimate for the surface area.

$$\text{Area (m}^2) = 0.202 \times W^{0.425} \times H^{0.725} \tag{11.1}$$

Here W is the weight of the person in kilograms, and H is the height of the person in meters.

The surface area of a 70-kg man of height 1.55 m is about 1.70 m^2. His metabolic rate at rest is therefore (40 Cal/m^2-hr) $\times$ 1.70 m^2 = 68 Cal/hr, or about 70 Cal/hr as stated in our earlier example. This metabolic rate at rest is called the *basal metabolic rate*.

11.2 Energy from Food

The chemical energy used by animals is obtained from the oxidation of food molecules. The glucose sugar molecule, for example, is oxidized as follows:

$$C_6H_{12}O_6 + 6O_2 \rightarrow 6CO_2 + 6H_2O + \text{energy} \tag{11.2}$$

For every gram of glucose ingested by the body, 3.81 Cal of energy is released for metabolic use.

The caloric value per unit weight is different for various foods. Measurements show that, on the average, carbohydrates (sugars and starches) and proteins provide about 4 Cal/g; lipids (fats) produce 9 Cal/g, and the oxidation of alcohol produces 7 Cal/g.[2]

The oxidation of food, which releases energy, does not occur spontaneously at normal environmental temperatures. For oxidation to proceed at body temperature, a catalyst must promote the reaction. In living systems, complex molecules, called enzymes, provide this function.

In the process of obtaining energy from food, oxygen is always consumed. It has been found that, independent of the type of food being utilized, 4.83 Cal of energy are produced for every liter of oxygen consumed. Knowing this relationship, one can measure with relatively simple techniques the metabolic rate for various activities (see Exercise 11-1).

The daily food requirements of a person depend on his or her activities. A sample schedule and the associated metabolic energy expenditure per square meter are shown in Table 11.2. Assuming, as before, that the surface area of the person whose activities are shown in Table 11.2 is 1.7 m^2, his/her total energy expenditure is 3940 Cal/day. If the person spent half the day sleeping and half the day resting in bed, the daily energy expenditure would be only 1530 Cal.

For most people the energy expenditure is balanced by the food intake. For example, the daily energy needs of the person whose activities are shown

[2]The high caloric content of alcohol presents a problem for people who drink heavily. The body utilizes fully the energy released by the oxidation of alcohol. Therefore, people who obtain a significant fraction of metabolic energy from this source reduce their intake of conventional foods. Unlike other foods, however, alcohol does not contain vitamins, minerals, and other substances necessary for proper functioning. As a result, chronic alcoholics often suffer from diseases brought about by nutritional deficiencies.

TABLE 11.2 ▶ One Day's Metabolic Energy Expenditure

Activity	Energy expenditure (Cal/m^2)
8 hr sleeping (35 Cal/m^2-hr)	280
8 hr moderate physical labor (150 Cal/m^2-hr)	1200
4 hr reading, writing, TV watching (60 Cal/m^2-hr)	240
1 hr heavy exercise (300 Cal/m^2-hr)	300
3 hr dressing, eating (100 Cal/m^2-hr)	300
Total expenditure	2320

in Table 11.2 (surface area 1.7 m^2) are met by the consumption of 400 g of carbohydrates, 200 g of protein, and 171 g of fat.

The composition and energy content of some common foods are shown in Table 11.3. Note that the sum of the weights of the protein, carbohydrates, and fat is smaller than the total weight of the food. The difference is due mostly to the water content of the food. The energy values quoted in the table reflect the fact that the caloric content of different proteins, carbohydrates, and fats deviate somewhat from the average values stated in the text.

If an excess of certain substances, such as water and salt, is ingested, the body is able to eliminate it. The body has no mechanism, however, for eliminating an excess in caloric intake. Over a period of time the excess energy is used by the body to manufacture additional tissue. If the consumption of excess food occurs simultaneously with heavy exercise, the energy may be utilized to increase the weight of the muscles. Most often, however, the excess

TABLE 11.3 ▶ Composition and Energy Content of Some Common Foods

Food	Total weight (g)	Protein weight (g)	Carbohydrate weight (g)	Fat weight (g)	Total energy (Cal)
Whole milk, 1 quart	976	32	48	40	660
Egg, 1	50	6	0	12	75
Hamburger, 1	85	21	0	17	245
Carrots, 1 cup	150	1	10	0	45
Potato (1 med., baked)	100	2	22	0	100
Apple	130	0	18	0	70
Bread, rye, 1 slice	23	2	12	0	55
Doughnut	33	2	17	7	135

energy is stored in fatty tissue that is manufactured by the body. Conversely, if the energy intake is lower than the demand, the body consumes its own tissue to make up the deficit. While the supply lasts, the body first utilizes its stored fat. For every 9 Cal of energy deficit, about 1 g of fat is used. Under severe starvation, once the fat is used up, the body begins to consume its own protein. Each gram of consumed protein yields about 4 Cal. Consumption of body protein results in the deterioration of body functions, of course. A relatively simple calculation (see Exercise 11-4) shows that an average healthy person can survive without food but with adequate water up to about 50 days. Overweight people can do better, of course. The "Guinness Book of World Records" states that Angus Barbieri of Scotland fasted from June, 1965, to July, 1966, consuming only tea, coffee, and water. During this period, his weight declined from 472 lb to 178 lb.

For a woman, the energy requirements increase somewhat during pregnancy due to the growth and metabolism of the fetus. As the following calculation indicates, the energy needed for the growth of the fetus is actually rather small. Let us assume that the weight gain of the fetus during the 270 days of gestation is uniform.[3] If at birth the fetus weighs 3 kg, each day it gains 11 g. Because 75% of tissue consists of water and inorganic minerals, only 2.75 g of the daily mass increase is due to organic materials, mainly protein. Therefore, the extra Calories per day required for the growth of the fetus is

$$\text{Calories required} = \frac{2.75 \text{ g protein}}{\text{day}} \times \frac{4 \text{ Cal}}{\text{g protein}} = 11 \text{ Cal/day}$$

To this number, we must add the basal metabolic consumption of the fetus. At birth, the surface area of the fetus is about 0.13 m^2 (from Eq. 11.1); therefore, at most, the basal metabolic consumption of the fetus per day is about $0.13 \times 40 \times 24 = 125$ Cal. Thus, the total increase in the energy requirement of a pregnant woman is only about $(125 + 11)$ Cal/day $= 136$ Cal/day. Actually, it may not even be necessary for a pregnant woman to increase her food intake, as the energy requirements of the fetus may be balanced by decreased physical activity during pregnancy. Various other aspects of metabolic energy balance are examined in Exercises 11-2 to 11-5.

11.3 Regulation of Body Temperature

People and other warm-blooded animals must maintain their body temperatures at a nearly constant level. For example, the normal internal body temperature of a person is about 37°C. A deviation of one or two degrees in

[3]This is a simplification because the weight gain is not uniform. It is greatest toward the end of gestation.

either direction may signal some abnormality. If the temperature-regulating mechanisms fail and the body temperature rises to 44° or 45°C, the protein structures are irreversibly damaged. A fall in body temperature below about 28°C results in heart stoppage.

The body temperature is sensed by specialized nerve centers in the brain and by receptors on the surface of the body. The various cooling or heating mechanisms of the body are then activated in accord with the temperature. The efficiency of muscles in performing external work is at best 20%. Therefore, at least 80% of the energy consumed in the performance of a physical activity is converted into heat inside the body. In addition, the energy consumed to maintain the basic metabolic processes is ultimately all converted to heat. If this heat were not eliminated, the body temperature would quickly rise to a dangerous level. For example, during moderate physical activity, a 70-kg man may consume 260 Cal/hr. Of this amount, at least 208 Cal is converted to heat. If this heat remained within the body, the body temperature would rise by 3 C°/hr. Two hours of such an activity would cause complete collapse. Fortunately, the body possesses a number of highly efficient methods for controlling the heat flow out of the body, thereby maintaining a stable internal temperature.

Most of the heat generated by the body is produced deep in the body, far from the surfaces. In order to be eliminated, this heat must first be conducted to the skin. For heat to flow from one region to another, there must be a temperature difference between the two regions. Therefore, the temperature of the skin must be lower than the internal body temperature. In a warm environment, the temperature of the human skin is about 35°C. In a cold environment, the temperature of some parts of the skin may drop to 27°C.

The tissue of the body, without blood flowing through it, is a poor conductor. Its thermal conductivity is comparable to that of cork (see Table 9.2). (K_c for tissue without blood is 18 Cal-cm/m^2-hr-C°.) Simple thermal conductivity through tissue is inadequate for elimination of the excess heat generated by the body. The following calculation illustrates this point. Assume that the thickness of the tissue between the interior and the exterior of the body is 3 cm and that the average area through which conduction can occur is 1.5 m^2. With a temperature difference T between the inner body and the skin of 2°C, the heat flow H per hour is, from Eq. 9.3,

$$H = \frac{K_c A \Delta T}{L} = \frac{18 \times 1.5 \times 2}{3} = 18 \text{ Cal/hr} \tag{11.3}$$

In order to increase the conductive heat flow to a moderate level of say 150 Cal/hr, the temperature difference between the interior body and the skin would have to increase to about 17 C°.

Fortunately the body possesses another method for transferring heat. Most of the heat is transported from the inside of the body by blood in the circulatory system. Heat enters the blood from an interior cell by conduction. In this case, heat transfer by conduction is relatively fast because the distances between the capillaries and the heat-producing cells are small. The circulatory system carries the heated blood near to the surface skin. The heat is then transferred to the outside surface by conduction. In addition to transporting heat from the interior of the body, the circulatory system controls the insulation thickness of the body. When the heat flow out of the body is excessive, the capillaries near the surface become constricted and the blood flow to the surface is greatly reduced. Because tissue without blood is a poor heat conductor, this procedure provides a heat-insulating layer around the inner body core.

11.4 Control of Skin Temperature

As was stated, for heat to flow out of the body, the temperature of the skin must be lower than the internal body temperature. Therefore, heat must be removed from the skin at a sufficient rate to ensure that this condition is maintained. Because the heat conductivity of air is very low (2.02 Cal-cm/m^2-hr-C°), if the air around the skin is confined—for example, by clothing—the amount of heat removed by conduction is small. The surface of the skin is cooled primarily by convection, radiation, and evaporation. However, if the skin is in contact with a good thermal conductor such as a metal, a considerable amount of heat can be removed by conduction (see Exercise 11-6).

11.5 Convection

When the skin is exposed to open air or some other fluid, heat is removed from it by convection currents. The rate of heat removal is proportional to the exposed surface area and to the temperature difference between the skin and the surrounding air. The rate of heat transfer by convection H_c' (see Eq. 9.4) is given by

$$H_c' = K_c' A_c (T_s - T_a) \tag{11.4}$$

where A_c is the skin area exposed to the open air; T_s and T_a are the skin and air temperatures, respectively; and K_c' is the convection coefficient, which has a value that depends primarily on the prevailing wind velocity. The value of K_c' as a function of air velocity is shown in Fig. 11.1. As the plot shows, the convection coefficient initially increases sharply with wind velocity, and then the increase becomes less steep (see Exercise 11-7).

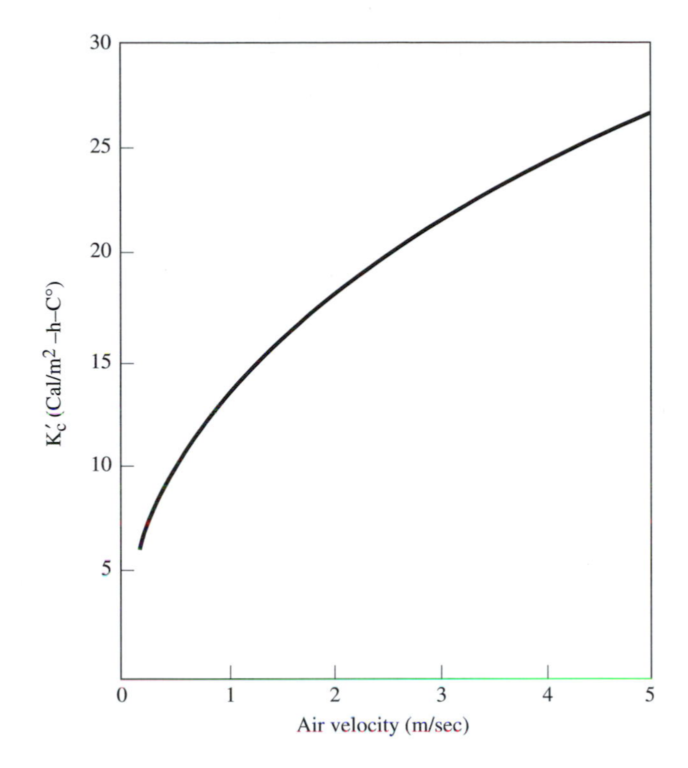

FIGURE 11.1 ▶ Convection coefficient as a function of air velocity.

The exposed area A_c is generally smaller than the total surface area of the body. For a naked person standing with legs together and arms close to the body, about 80% of the surface area is exposed to convective air currents. (The exposed area can be reduced by curling up the body.)

Note that heat flows from the skin to the environment only if the air is colder than the skin. If the opposite is the case, the skin is actually heated by the convective air flow.

Let us now calculate the amount of heat removed from the skin by convection. Consider a naked person whose total surface area is 1.7 m^2. Standing straight, the exposed area is about 1.36 m^2. If the air temperature is 25°C and the average skin temperature is 33°C, the amount of heat removed is

$$H_c' = 1.36 K_c' \times 8 = 10.9 K_c' \text{ Cal/hr}$$

Under nearly windless conditions, K_c' is about 6 Cal/m^2-hr-C° (see Fig. 11.1), and the convective heat loss is 65.4 Cal/hr. During moderate work, the energy consumption for a person of this size is about 170 Cal/hr. Clearly, convection in a windless environment does not provide adequate cooling. The wind velocity has to increase to about 1.5 m/sec to provide cooling at a rate of 170 Cal/hr.

11.6 Radiation

Equation 9.6 shows that the energy exchange by radiation H_r involves the fourth power of temperature; that is,

$$H_r = e\sigma\left(T_1^4 - T_2^4\right)$$

However, because in the environment encountered by living systems the temperature on the absolute scale seldom varies by more than 15%, it is possible to use, without much error, a linear expression for the radiative energy exchange (see Exercise 11-8a and b); that is,

$$H_r = K_r A_r e(T_s - T_r) \tag{11.5}$$

where T_s and T_r are the skin surface temperature and the temperature of the nearby radiating surface, respectively; A_r is the area of the body participating in the radiation; e is the emissivity of the surface; and K_r is the radiation coefficient. Over a fairly wide range of temperatures, K_r is, on the average, about 6.0 Cal/m^2-hr-C° (see Exercise 11-8c).

The environmental radiating surface and skin temperatures are such that the wavelength of the thermal radiation is predominantly in the infrared region of the spectrum. The emissivity of the skin in this wavelength range is nearly unity, independent of the skin pigmentation. For a person with $A_r = 1.5$ m^2, $T_r = 25$°C and $T_s = 32$°C. The radiative heat loss is 63 Cal/hr.

If the radiating surface is warmer than the skin surface, the skin is heated by radiation. A person begins to feel discomfort due to radiation if the temperature difference between the exposed skin and the radiating environment exceeds about 6 C°. In the extreme case, when the skin is illuminated by the sun or some other very hot object like a fire, the skin is heated intensely. Because the temperature of the source is now much higher than the temperature of the skin, the simplified expression in Eq. 11.5 no longer applies.

11.7 Radiative Heating by the Sun

The intensity of solar energy at the top of the atmosphere is about 1150 Cal/m^2-hr. Not all this energy reaches the surface of the Earth. Some of it is reflected by airborne particles and water vapor. A thick cloud cover may reflect as much as 75% of solar radiation. The inclination of the Earth's axis of rotation

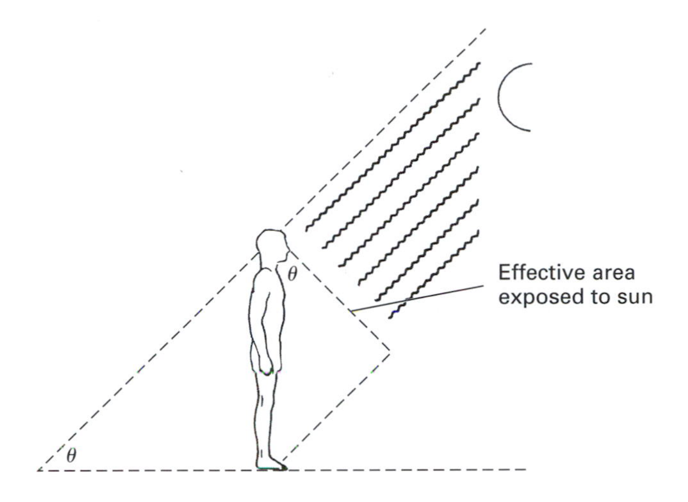

FIGURE 11.2 ▶ Radiative heating by the sun.

further reduces the intensity of solar radiation at the surface. However, in dry equatorial deserts, nearly all the solar radiation may reach the surface.

Because the rays of the sun come from one direction only, at most half the body surface is exposed to solar radiation. In addition the area perpendicular to the solar flux is reduced by the cosine of the angle of incidence (see Fig. 11.2). As the sun approaches the horizon, the effective area for the interception of radiation increases, but at the same time the radiation intensity decreases because the radiation passes through a thicker layer of air. Still, the amount of solar energy heating the skin can be very large. Assuming that the full intensity of solar radiation reaches the surface, the amount of heat H_r that the human body receives from solar radiation is

$$H_r = 1150/2 \times e \times A \cos\theta \text{ Cal/hr} \tag{11.6}$$

Here A is the skin area of the person, θ is the angle of incidence of sunlight, and e is the emissivity of the skin. The emissivity of the skin in the wavelength region of solar radiation depends on the pigmentation. Dark skin absorbs about 80% of the radiation, and light skin absorbs about 60%. From Eq. 11.6, a light-skinned person with a skin area of 1.7 m^2, subject to intense solar radiation incident at a 60° angle, receives heat at the rate of 294 Cal/hr. Radiative heating is decreased by about 40% if the person wears light-colored clothing. Radiative heating is also reduced by changing the orientation of the body with respect to the sun. Camels resting in the shadeless desert face the sun, which minimizes the skin area exposed to solar radiation.

11.8 Evaporation

In a warm climate, convection and radiation cannot adequately cool a person engaged in even moderate physical activity. A large fraction of cooling is provided by the evaporation of sweat from the skin surfaces. At normal skin temperatures, the latent heat of vaporization for water is 0.580 Cal/g. Therefore, about 580 Cal of heat are removed for each liter of sweat that evaporates from the skin. The body contains two types of sweat glands, the eccrine and the apocrine. The eccrine glands are distributed over the whole surface of the body, and they respond primarily to the nerve impulses generated by the thermoregulatory system of the body. As the heat load on the body rises, the sweat secreted by these glands increases proportionately. There is an exception to this. The *eccrine glands* in the palms of the hand and the soles of the feet are stimulated by elevated levels of adrenaline in the blood, which may result from emotional stress.

The *apocrine* sweat glands, found mostly in the pubic regions, are not associated with temperature control. They are stimulated by adrenaline in the blood stream, and they secrete a sweat rich in organic matter. The decomposition of these substances produces body odor.

The ability of the human body to secrete sweat is remarkable. is remarkable. For brief periods of time, a person can produce sweat at a rate up to 4 liter/hr. Such a high rate of sweating, however, cannot be maintained. For longer periods, up to 6 hours, a sweating rate of 1 liter/hr is common in the performance of heavy work in a hot environment.

During prolonged heavy sweating, adequate amounts of water must be drunk; otherwise, the body becomes dehydrated. A person's functioning is severely limited when dehydration results in a 10% loss of body weight. Some desert animals can endure greater dehydration than humans; a camel, for example, may lose water amounting to 30% of its body weight without serious consequences.

Only sweat that evaporates is useful in cooling the skin. Sweat that rolls off or is wiped off does not provide significant cooling. Nevertheless, excess sweat does ensure full wetting of the skin. The amount of sweat that evaporates from the skin depends on ambient temperature, humidity, and air velocity. Evaporative cooling is most efficient in a hot, dry, windy environment.

There is another avenue for evaporative heat loss: breathing. The air leaving the lungs is saturated by water vapor from the moist lining of the respiratory system. At a normal human breathing rate, the amount of heat removed by this avenue is small, less than 9 Cal/hr (see Exercise 11-9); however, for furred animals that do not sweat, this method of heat removal is very important. These animals can increase heat loss by taking short shallow breaths (panting) that do not bring excessive oxygen into the lungs but do pick up moisture from the upper respiratory tract.

By evaporative cooling, a person can cope with the heat generated by moderate activity even in a very hot, sunny environment. To illustrate this, we will calculate the rate of sweating required for a person walking nude in the sun at a rate of 3 mph, with the ambient temperature at 47°C (116.6°F).

With a skin area of 1.7 m^2, the energy consumed in the act of walking is about 240 Cal/hr. Almost all this energy is converted to heat and delivered to the skin. In addition, the skin is heated by convection and by radiation from the environment and the sun. The heat delivered to the skin by convection is

$$H_c' = K_c' A_c (T_s - T_a)$$

For a 1-m/sec wind, K_c' is 13 Cal/m^2-hr-C°. The exposed area A_c is about 1.5 m^2. If the skin temperature is 36°C,

$$H_c' = 13 \times 1.50 \times (47 - 36) = 215 \text{ Cal/hr}$$

As calculated previously, the radiative heating by the sun is about 294 Cal/hr. The radiative heating by the environment is

$$H_c' = K_r A_r e (T_r - T_s) = 6 \times 1.5 \times (47 - 36) = 99 \text{ Cal/hr}$$

In this example, the only mechanism available for cooling the body is the evaporation of sweat. The total amount of heat that must be removed is (240 + 215 + 194 + 99) Cal/hr = 848 Cal/hr. The evaporation of about 1.5 liter/hr of sweat will provide the necessary cooling. Of course, if the person is protected by light clothing, the heat load is significantly reduced. The human body is indeed very well equipped to withstand heat. In controlled experiments, people have survived a temperature of 125°C for a period of time that was adequate to cook a steak.

11.9 Resistance to Cold

In a thermally comfortable environment, the body functions at a minimum expenditure of energy. As the environment cools, a point is reached where the basal metabolic rate increases to maintain the body temperature at a proper level. The temperature at which this occurs is called the *critical temperature*. This temperature is a measure of the ability of an animal to withstand cold.

Human beings are basically tropical animals. Unprotected, they are much better able to cope with heat than with cold. The critical temperature for humans is about 30°C. By contrast, the critical temperature for the heavily furred arctic fox is −40°C.

The discomfort caused by cold is due primarily to the increased rate of heat outflow from the skin. This rate depends not only on the temperature but also on the wind velocity and humidity. For example, at 20°C, air moving with a velocity of 30 cm/sec removes more heat than still air at 15°C. In this case, a mild wind at 30 cm/sec is equivalent to a temperature drop of more than 5 C°.

The body defends itself against cold by decreasing the heat outflow and by increasing the production of heat. When the temperature of the body begins to drop, the capillaries leading to the skin become constricted, reducing the blood flow to the skin. This results in a thicker thermal insulation of the body. In a naked person, this mechanism is fully utilized when the ambient temperature drops to about 19°C. At this point, the natural insulation cannot be increased any more.

Additional heat required to maintain the body temperature is obtained by increasing the metabolism. One involuntary response that achieves this is shivering. As shown in Table 11.1, shivering raises the metabolism to about 250 Cal/m^2-hr. If these defenses fail and the temperature of the skin and underlying tissue fall below about 5°C, frostbite and eventually more serious freezing occur.

The most effective protection against cold is provided by thick fur, feathers, or appropriate clothing. At −40°C, without insulation, the heat loss is primarily convective and radiative. By convection alone in moderately moving air, the rate of heat removal per square meter of skin surface is about 660 Cal/m^2-hr (see Exercise 11-10). With a thick layer of fur or similar insulation the skin is shielded from convection and the heat is transferred to the environment by conduction only. The thermal conductivity of insulating materials such as fur or down is K_c = 0.36 Cal cm/m^2-hr-C°; therefore, the heat transfer from the skin at 30°C to the ambient environment at −40°C through 1 cm of insulation is, from Eq. 9.3, 25.2 Cal/m^2-hr. This is below the basal metabolic rate for most animals. Although body heat is lost also through radiation and evaporation, our calculation indicates that well-insulated animals, including a clothed person, can survive in cold environments.

As stated earlier, at moderate temperatures the amount of heat removed by breathing at a normal rate is small. At very cold temperatures, however, the heat removed by this channel is appreciable. Although the heat removed by the evaporation of moisture from the lungs remains approximately constant, the amount of heat required to warm the inspired air to body temperature increases as the ambient air temperature drops. For a person at an ambient temperature of −40°C, the amount of heat removed from the body in the process of breathing is about 14.4 Cal/hr (see Exercise 11-11). For a well-insulated animal, this heat loss ultimately limits its ability to withstand cold.

11.10 Heat and Soil

Much of life depends directly or indirectly on biological activities near the surface of the soil. In addition to plants, there are worms and insects whose lives are soil-bound (1 acre of soil may contain 500 kg of earthworms). Soil is also rich in tiny organisms such as bacteria, mites, and fungi whose metabolic activities are indispensable for the fertility of the soil. To all this life, the temperature of the soil is of vital importance.

The surface soil is heated primarily by solar radiation. Although some heat is conducted to the surface from the molten core of the Earth, the amount from this source is negligible compared to solar heating. The Earth is cooled by convection, radiation, and the evaporation of soil moisture. On the average, over a period of a year, the heating and cooling are balanced; therefore, over this period of time, the average temperature of the soil does not change appreciably. However, over shorter periods of time, from night to day, from winter to summer, the temperature of the top soil changes considerably; these fluctuations govern the life cycles in the soil.

The variations in soil temperature are determined by the intensity of solar radiation, the composition and moisture content of the soil, the vegetation cover, and atmospheric conditions such as clouds, wind, and airborne particles (see Exercises 11-12 and 11-13). Certain patterns, however, are general. During the day while the sun is shining, more heat is delivered to the soil than is removed by the various cooling mechanisms. The temperature of the soil surface therefore rises during the day. In dry soil, the surface temperature may increase by 3 or 4 C°/hr. The surface heating is especially intense in dry, unshaded deserts. Some insects living in these areas have evolved long legs to keep them removed from the hot surface.

The heat that enters the surface is conducted deeper into the soil. It takes some time, however, for the heat to propagate through the soil. Measurements show that a temperature change at the surface propagates into the soil at a rate of about 2 cm/hr. At night, the heat loss predominates, and the soil surface cools. The heat that was stored in the soil during the day now propagates to the surface and leaves the soil. Because of the finite time required for the heat to propagate through the soil, the temperature a few centimeters below the surface may still be rising while the surface is already cooling off. Some animals take advantage of this lag in temperature between the surface and the interior of the soil. They burrow into the ground to avoid the larger temperature fluctuations at the surface.

At the usual temperatures the thermal radiation emitted by the soil is in the infrared region of the spectrum, which is strongly reflected by water vapor and clouds. As a result, on cloudy days the thermal radiation emitted by the

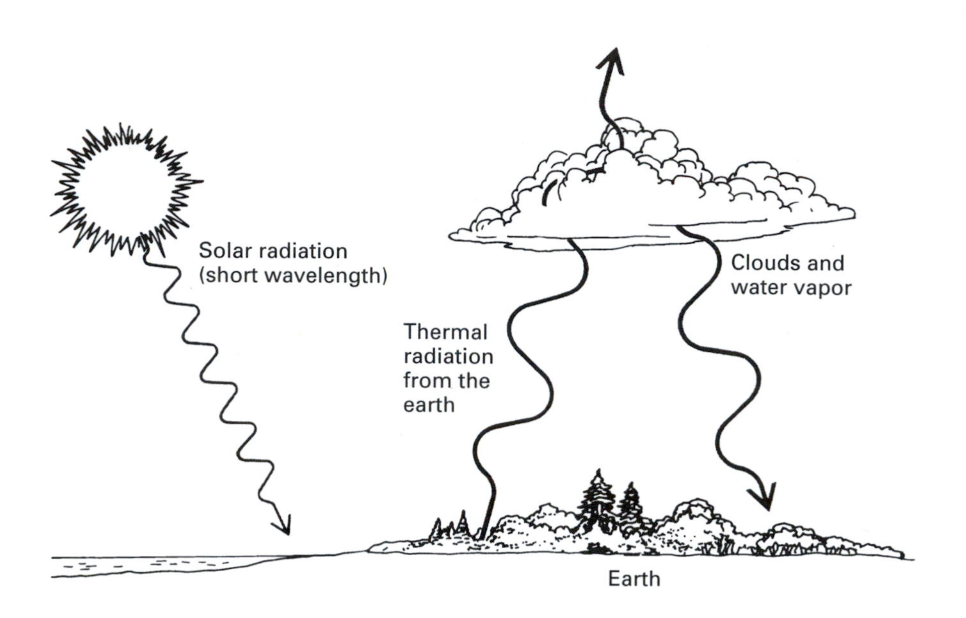

FIGURE 11.3 ▶ The greenhouse effect.

soil is reflected back, and the net outflow of heat from the soil is reduced—this is called the *greenhouse effect* (see Fig. 11.3).

▶ EXERCISES ▶

11-1. Design an experiment that would measure the metabolic rate of walking at 5 km/hr up a 20° slope.

11-2. How long can a man survive in an airtight room that has a volume of 27 m^3. Assume that his surface area is 1.70 m^2. Use data provided in the text.

11-3. A submarine carries an oxygen tank that holds oxygen at a pressure of 100 atm. What must be the volume of the tank to provide adequate oxygen for 50 people for 10 days? Assume that daily energy expenditure is as given in Table 11.2 and the average surface area of each person is 1.70 m^2.

11-4. Calculate the length of time that a person can survive without food but with adequate water. Obtain a solution under the following assumptions: (a) The initial weight and surface area of the person are 70 kg and 1.70 m^2, respectively. (b) The survival limit is reached when the person loses one-half his or her body weight. (c) Initially the body contains 5 kg of fatty tissue. (d) During the fast the person sleeps 8 hr/day and rests quietly the remainder of the time. (e) As the person loses weight, his or her surface area decreases (see Eq. 11.1). However, here we assume that the surface area remains unchanged.

11-5. Suppose that a person of weight 60 kg and height 1.4 m reduces her sleep by 1 hr/day and spends this extra time reading while sitting upright. If her food intake remains unchanged, how much weight will she lose in one year?

11-6. Assume that a person is sitting naked on an aluminum chair with 400-cm^2 area of the skin in contact with aluminum. If the skin temperature is 38 °C and the aluminum is kept at 25 °C, compute the amount of heat transfer per hour from the skin. Assume that the body contacting the aluminum is insulated by a layer of unperfused fat tissue 0.5 cm thick ($K_c = 18$ Cal cm/m^2-hr-C°) and that the heat conductivity of aluminum is very large. Is this heat transfer significant in terms of the metabolic heat consumption?

11-7. Explain qualitatively the functional dependence of K_c' on the air velocity (see Fig. 11.1).

11-8. (a) Show that $(T_s^4 - T_r^4) = (T_s^3 + T_s^2 T_r + T_s T_r^2 + T_r^3)(T_s - T_r)$. (b) Compute percentage change in the term $(T_s^3 + T_s^2 T_r + T_s T_r^2 + T_r^3)$ as the radiative temperature of the environment changes from 0 to 40 °C. (Note that the temperatures in the computations must be expressed on the absolute scale. However, if the expression contains only the difference between two temperatures, either the absolute or the centigrade scale may be used.) (c) Calculate the value of K_r in Eq. 11.5 under conditions discussed in the text where $T_r = 25$ °C (298 K), $T_s = 32$ °C (305 K), and $H_r = 63$ Cal/hr.

11-9. A person takes about 20 breaths per minute with 0.5 liter of air in each breath. How much heat is removed per hour by the moisture in the exhaled breath if the incoming air is dry and the exhaled breath is fully saturated? Assume that the water vapor pressure in the saturated exhaled air is 24 torr.

11-10. Compute the heat loss per square meter of skin surface at −40 °C in moderate wind (about 0.5 m/sec, $K_c' = 10$ Cal/m^2-hr-°C). Assume that the skin temperature is 26 °C.

11-11. Calculate the amount of heat required per hour to raise the temperature of inspired air from −40 °C to the body temperature of 37°C. Assume that the breathing rate is 600 liters of air per hour. (This is the breathing rate specified in Exercise 11-9.) The amount of heat required to raise the temperature of 1 mole of air (22.4 liter) by 1 C° at 1 atm is 29.2 J (6.98×10^{-3} Cal).

11-12. Explain why the daily temperature fluctuations in the soil are smaller (a) in wet soil than in dry soil, (b) in soil with a grass growth than in bare soil, and (c) when the air humidity is high.

11-13. Explain why the temperature drops rapidly at night in a desert.

11-14. The therapeutic effects of heat have been known since ancient times. Local heating, for example, relieves muscle pain and arthritic conditions. Discuss some effects of heat on tissue that may explain its therapeutic value.

Chapter 12

Waves and Sound

Most of the information about our physical surroundings comes to us through our senses of hearing and sight. In both cases we obtain information about objects without being in physical contact with them. The information is transmitted to us in the first case by sound, in the second case by light. Although sound and light are very different phenomena, they are both waves. A wave can be defined as a disturbance that carries energy from one place to another without a transfer of mass. The energy carried by the waves stimulates our sensory mechanisms.

In this chapter, we will first explain briefly the nature of sound and then review some general properties of wave motion applicable to both sound and light. Using this background we will examine the process of hearing and some other biological aspects of sound. Light will be discussed in Chapter 15.

12.1 Properties of Sound

Sound is a mechanical wave produced by vibrating bodies. For example, when an object such as a tuning fork or the human vocal cords is set into vibrational motion, the surrounding air molecules are disturbed and are forced to follow the motion of the vibrating body. The vibrating molecules in turn transfer their motion to adjacent molecules causing the vibrational disturbance to propagate away from the source. When the air vibrations reach the ear, they cause the eardrum to vibrate; this produces nerve impulses that are interpreted by the brain.

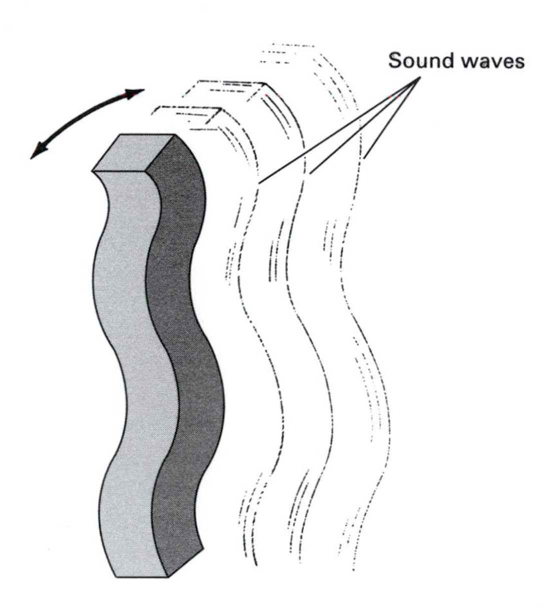

FIGURE 12.1 ▶ A complex vibrational pattern.

All matter transmits sound to some extent, but a material medium is needed between the source and the receiver to propagate sound. This is demonstrated by the well-known experiment of the bell in the jar. When the bell is set in motion, its sound is clearly audible. As the air is evacuated from the jar, the sound of the bell diminishes and finally the bell becomes inaudible.

The propagating disturbance in the sound-conducting medium is in the form of alternate compressions and rarefactions of the medium, which are initially caused by the vibrating sound source. These compressions and rarefactions are simply deviations in the density of the medium from the average value. In a gas, the variations in density are equivalent to pressure changes.

Two important characteristics of sound are *intensity*, which is determined by the magnitude of compression and rarefaction in the propagating medium, and *frequency*, which is determined by how often the compressions and rarefactions take place. Frequency is measured in cycles per second, which is designated by the unit *hertz* after the scientist Heinrich Hertz. The symbol for this unit is Hz. (1 Hz = 1 cycle per second.)

The vibrational motion of objects can be highly complex (see Fig. 12.1), resulting in a complicated sound pattern. Still, it is useful to analyze the properties of sound in terms of simple sinusoidal vibrations such as would be set up by a vibrating tuning fork (see Fig. 12.2). The type of simple sound

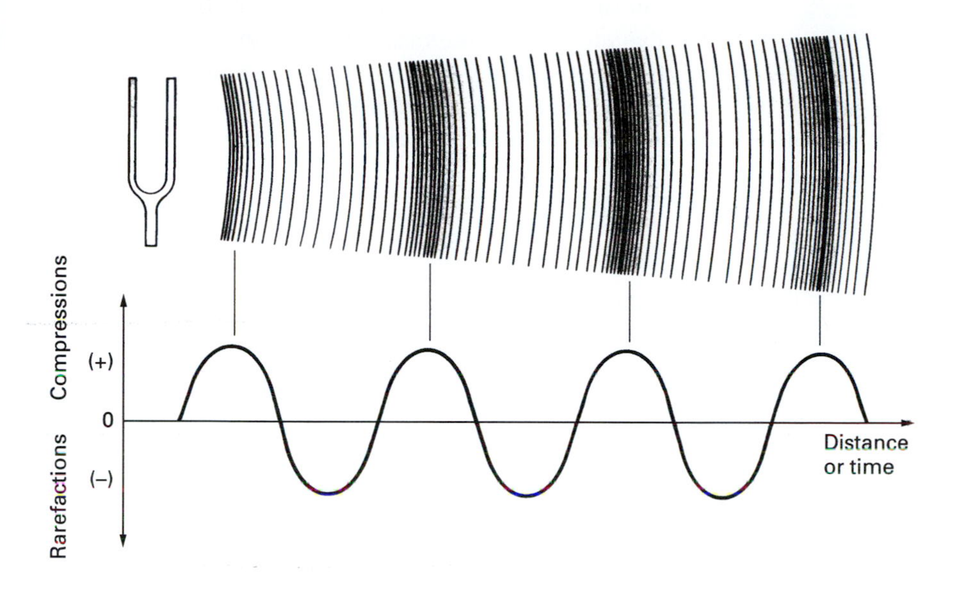

FIGURE 12.2 ▶ Sinusoidal sound wave produced by a vibrating tuning fork.

pattern shown in Fig. 12.2 is called a *pure tone*. When a pure tone propagates through air, the pressure variations due to the compressions and rarefactions are sinusoidal in form.

If we were to take a "snapshot" of the sound at a given instant in time, we would see pressure variations in space, which are also sinusoidal. (Such pictures can actually be obtained with special techniques.) In such a picture the distance between the nearest equal points on the sound wave is called the *wavelength* λ.

The speed of the sound wave v depends on the material that propagates the sound. In air at 20°C, the speed of sound is about 3.3×10^4 cm/sec, and in water it is about 1.4×10^5 cm/sec. In general, the relationship between frequency, wavelength, and the speed of propagation is given by the following equation:

$$v = \lambda f \tag{12.1}$$

This relationship between frequency, wavelength, and speed is true for all types of wave motions.

The pressure variations due to the propagating sound are superimposed on the ambient air pressure. Thus, the total pressure in the path of a sinusoidal sound wave is of the form

$$P = P_a + P_o \sin 2\pi f t \tag{12.2}$$

where P_a is the ambient air pressure (which at sea level at 0°C is 1.01×10^5Pa $= 1.01 \times 10^6$ dyn/cm^2), P_o is the maximum pressure change due to the sound wave, and f is the frequency of the sound. The amount of energy transmitted by a sinusoidal sound wave per unit time through each unit area perpendicular to the direction of sound propagation is called the *intensity* I and is given by

$$I = \frac{P_o^2}{2\rho v} \tag{12.3}$$

Here ρ is the density of the medium, and v is the speed of sound propagation.

12.2 Some Properties of Waves

All waves, including sound and light, exhibit the phenomena of reflection, refraction, interference, and diffraction. These phenomena, which play an important role in both hearing and seeing, are described in detail in most basic physics texts (see [12-7]). Here we will review them only briefly.

12.2.1 Reflection and Refraction

When a wave enters one medium from another, part of the wave is reflected at the interface, and part of it enters the medium. If the interface between the two media is smooth on the scale of the wavelength (i.e., the irregularities of the interface surface are smaller than λ), the reflection is specular (mirrorlike). If the surface has irregularities that are larger than the wavelength, the reflection is diffuse. An example of diffuse reflection is light reflected from paper.

If the wave is incident on the interface at an oblique angle, the direction of propagation of the transmitted wave in the new medium is changed (see Fig. 12.3). This phenomenon is called *refraction*. The angle of reflection is always equal to the angle of incidence, but the angle of the refracted wave is, in general, a function of the properties of the two media. The fraction of the energy transmitted from one medium to another depends again on the properties of the media and on the angle of incidence. For a sound wave incident perpendicular to the interface, the ratio of transmitted to incident intensity is given by

$$\frac{I_t}{I_i} = \frac{4\rho_1 v_1 \rho_2 v_2}{(\rho_1 v_1 + \rho_2 v_2)^2} \tag{12.4}$$

where the subscripted quantities are the velocity and density in the two media. The solution of Eq. 12.4 shows that when sound traveling in air is incident perpendicular to a water surface, only about 0.1% of the sound energy enters the

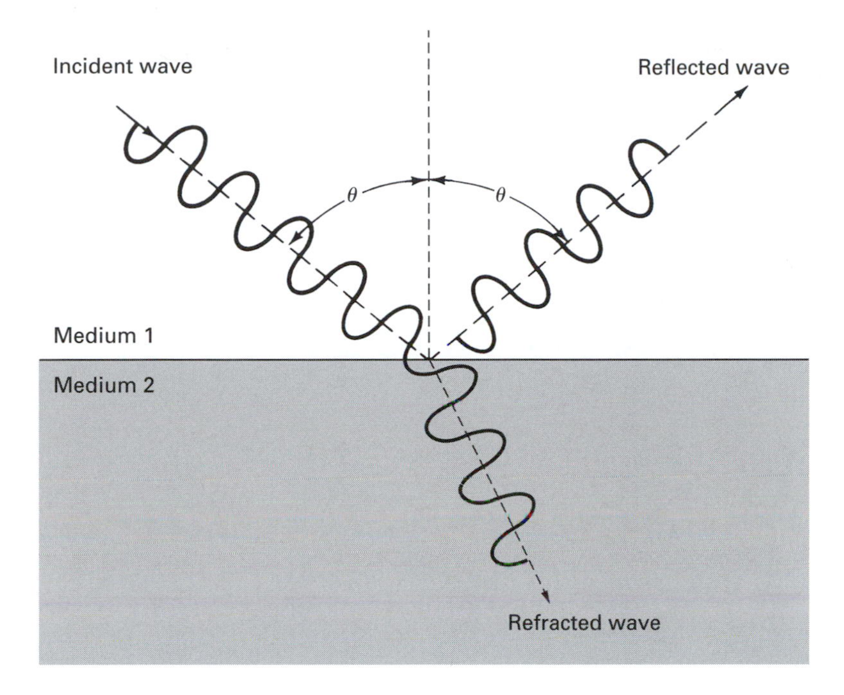

FIGURE 12.3 ▶ Illustration of reflection and refraction. (θ is the angle of incidence.)

water; 99.9% is reflected. The fraction of sound energy entering the water is even smaller when the angle of incidence is oblique. Water is thus an efficient barrier to sound.

12.2.2 Interference

When two (or more) waves travel simultaneously in the same medium, the total disturbance in the medium is at each point the vectorial sum of the individual disturbances produced by each wave. This phenomenon is called *interference*. For example, if two waves are in phase, they add so that the wave disturbance at each point in space is increased. This is called *constructive interference* (see Fig. 12.4a). If two waves are out of phase by 180°, the wave disturbance in the propagating medium is reduced. This is called *destructive interference* (Fig. 12.4b). If the magnitudes of two out-of-phase waves are the same, the wave disturbance is completely canceled (Fig. 12.4c).

A special type of interference is produced by two waves of the same frequency and magnitude traveling in opposite directions. The resultant wave pattern is stationary in space and is called a *standing wave*. Such standing

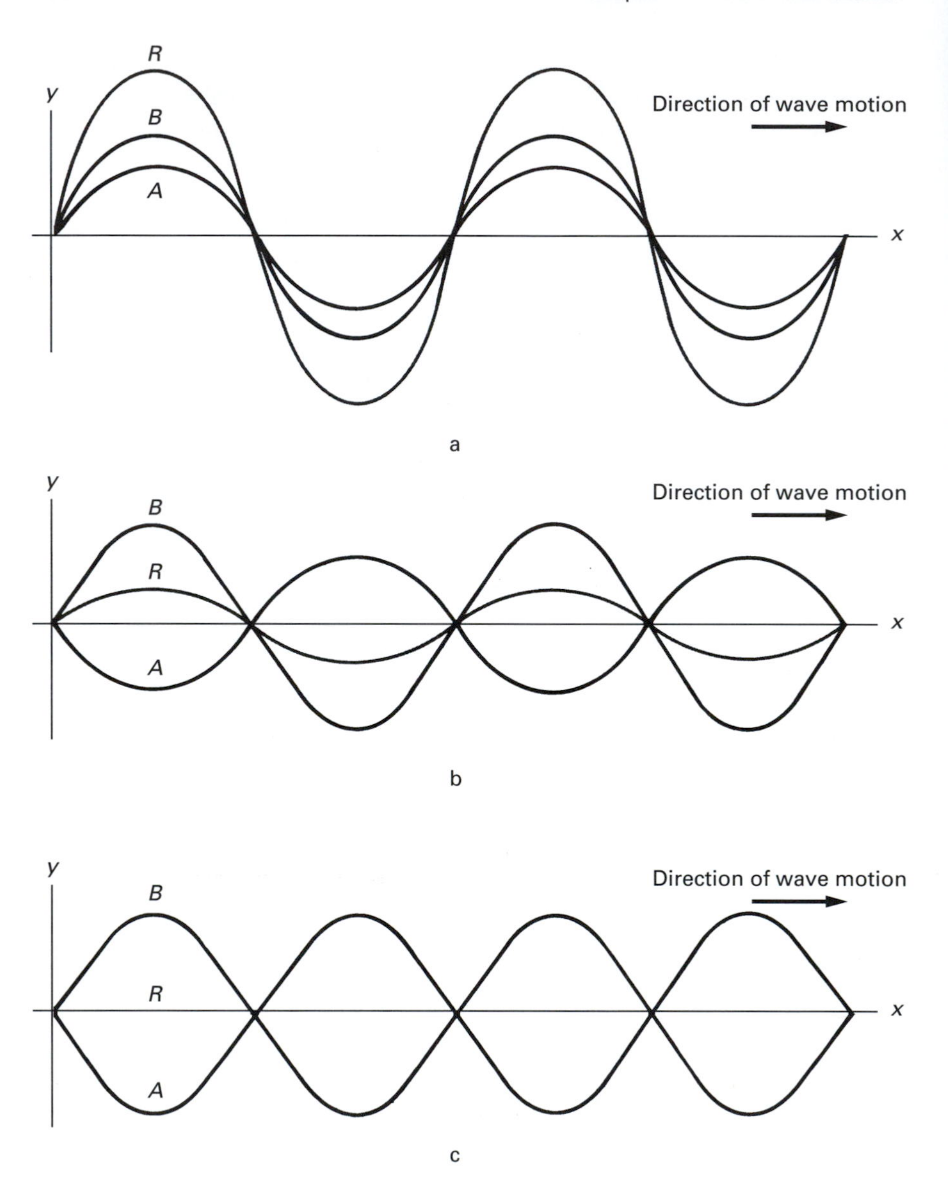

FIGURE 12.4 ▶ (a) Constructive interference. (b, c) Destructive interference. *R* is the resultant of the interference of the two waves *A* and *B*.

sound waves are formed in hollow pipes such as the flute. It can be shown that, in a given structure, standing waves can exist only at specific frequencies, which are called *resonant frequencies*.

12.2.3 Diffraction

Waves have a tendency to spread as they propagate through a medium. As a result, when a wave encounters an obstacle, it spreads into the region behind the obstacle. This phenomenon is called *diffraction*. The amount of diffraction depends on the wavelength: The longer the wavelength, the greater is the spreading of the wave. Significant diffraction into the region behind the obstacle occurs only if the size of the obstacle is smaller than the wavelength. For example, a person sitting behind a pillar in an auditorium hears the performer because the long wavelength sound waves spread behind the pillar. But the view of the performance is obstructed because the wavelength of light is much smaller than the pillar, and, therefore, the light does not diffract into the region behind the pillar.

Objects that are smaller than the wavelength do not produce a significant reflection. This too is due to diffraction. The wave simply diffracts around the small obstacle, much as flowing water spreads around a small stick.

Both light waves and sound waves can be focused with curved reflectors and lenses. There is, however, a limit to the size of the focused spot. It can be shown that the diameter of the focused spot cannot be smaller than about $\lambda/2$. These properties of waves have important consequences in the process of hearing and seeing.

12.3 Hearing and the Ear

The sensation of hearing is produced by the response of the nerves in the ear to pressure variations in the sound wave. The nerves in the ear are not the only ones that respond to pressure, as most of the skin contains nerves that are pressure-sensitive. However, the ear is much more sensitive to pressure variations than any other part of the body.

Figure 12.5 is a drawing of the human ear. (The ear construction of other terrestrial vertebrates is similar.) For the purposes of description, the ear is usually divided into three main sections: the outer ear, the middle ear, and the inner ear. The sensory cells that convert sound to nerve impulses are located in the liquid-filled inner ear.

The main purpose of the outer and middle ears is to conduct the sound into the inner ear.

The outer ear is composed of an external flap called the *pinna* and the ear canal, which is terminated by the *tympanic membrane* (eardrum). In many animals the pinna is large and can be rotated toward the source of the sound; this helps the animal to locate the source of sound. However, in humans the pinna is fixed and so small that it does not seem to contribute significantly to the hearing process.

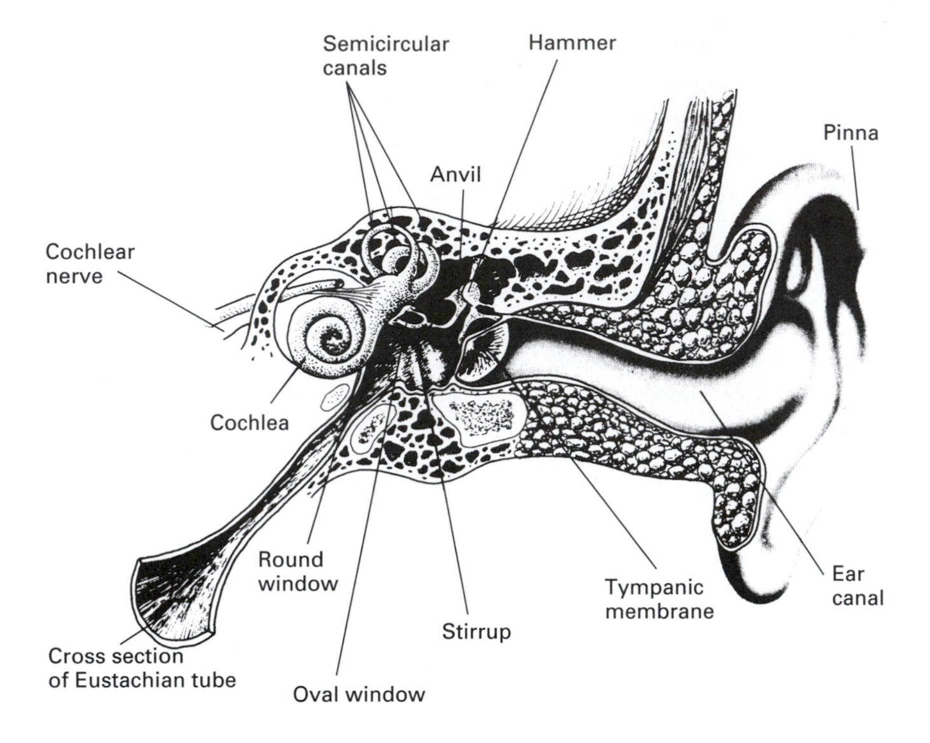

FIGURE 12.5 ▶ A semidiagrammatic drawing of the ear with various structures cut away and simplified to show the basic relationships more clearly. The middle ear muscles have been omitted.

The ear canal of an average adult is about 0.75 cm in diameter and 2.5 cm long, a configuration that is resonant for sound waves at frequencies around 3000 Hz. This accounts in part for the high sensitivity of the ear to sound waves in this frequency range.

For an animal to perceive sound, the sound has to be coupled from air to the sensory cells that are in the fluid environment of the inner ear. We showed earlier that direct coupling of sound waves into a fluid is inefficient because most of the sound energy is reflected at the interface. The middle ear provides an efficient conduction path for the sound waves from air into the fluid of the inner ear.

The middle ear is an air-filled cavity that contains a linkage of three bones called *ossicles* that connect the eardrum to the inner ear. The three bones are called the *hammer*, the *anvil*, and the *stirrup*. The hammer is attached to the inner surface of the eardrum, and the stirrup is connected to the oval window, which is a membrane-covered opening in the inner ear.

When sound waves produce vibrations in the eardrum, the vibrations are transmitted by the ossicles to the oval window, which in turn sets up pressure variations in the fluid of the inner ear. The ossicles are connected to the walls of the middle ear by muscles that also act as a volume control. If the sound is excessively loud, these muscles as well as the muscles around the eardrum stiffen and reduce the transmission of sound to the inner ear.

The middle ear serves yet another purpose. It isolates the inner ear from the disturbances produced by movements of the head, chewing, and the internal vibrations produced by the person's own voice. To be sure, some of the vibrations of the vocal cords are transmitted through the bones into the inner ear, but the sound is greatly attenuated. We hear ourselves talk mostly by the sound reaching our eardrums from the outside. This can be illustrated by talking with the ears plugged.

The *Eustachian tube* connects the middle ear to the upper part of the throat. Air seeps in through this tube to maintain the middle ear at atmospheric pressure. The movement of air through the Eustachian tube is aided by swallowing. A rapid change in the external air pressure such as may occur during an airplane flight causes a pressure imbalance on the two sides of the eardrum. The resulting force on the eardrum produces a painful sensation that lasts until the pressure in the middle ear is adjusted to the external pressure. The pain is especially severe and prolonged if the Eustachian tube is blocked by swelling or infection.

The conversion of sound waves into nerve impulses occurs in the *cochlea*, which is located in the inner ear. The cochlea is a spiral cavity shaped like a snail shell. The wide end of the cochlea, which contains the oval and the round windows, has an area of about 4 mm^2. The cochlea is formed into a spiral with about $2\frac{3}{4}$ turns. If the cochlea were uncoiled, its length would be about 35 mm.

Inside the cochlea there are three parallel ducts; these are shown in the highly simplified drawing of the uncoiled cochlea in Fig. 12.6. All three ducts are filled with a fluid. The vestibular and tympanic canals are joined at the apex of the cochlea by a narrow opening called the *helicotrema*. The cochlear duct is isolated from the two canals by membranes. One of these membranes, called the *basilar membrane*, supports the auditory nerves.

The vibrations of the oval window set up a sound wave in the fluid filling the vestibular canal. The sound wave, which travels along the vestibular canal and through the helicotrema into the tympanic canal, produces vibrations in the basilar membrane which stimulate the auditory nerves to transmit electrical pulses to the brain (see Chapter 13). The excess energy in the sound wave is dissipated by the motion of the round window at the end of the tympanic canal.

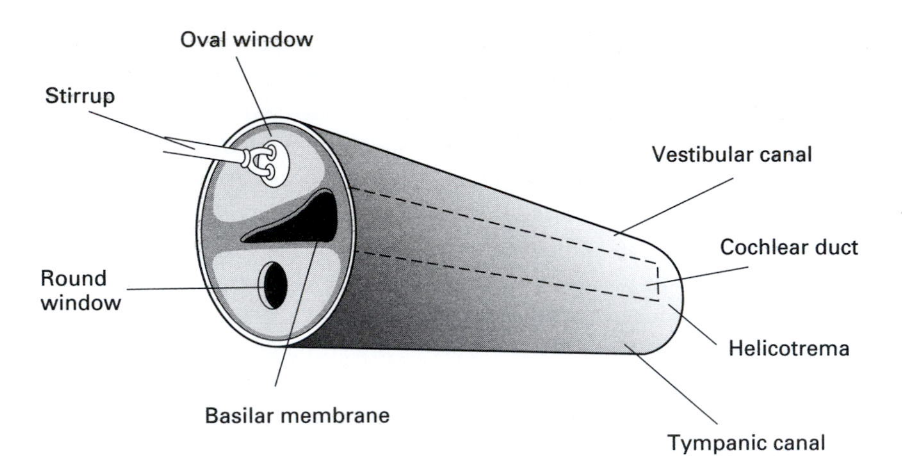

FIGURE 12.6 ► An uncoiled view of the cochlea.

12.3.1 Performance of the Ear

The nerve impulses evoke in the brain the subjective sensation of sound. *Loudness*, *pitch*, and *quality* are some of the terms we use to describe the sounds we hear. It is a great challenge for physiologists to relate these subjective responses with the physical properties of sound such as intensity and frequency. Some of these relationships are now well understood; others are still subjects for research.

In most cases, the sound wave patterns produced by instruments and voices are highly complex. Each sound has its own characteristic pattern. It would be impossible to evaluate the effect of sound waves on the human auditory system if the response to each sound pattern had to be analyzed separately. Fortunately the problem is not that complicated. About 150 years ago, J. B. J. Fourier, a French mathematician, showed that complex wave shapes can be analyzed into simple sinusoidal waves of different frequencies. In other words, a complex wave pattern can be constructed by adding together a sufficient number of sinusoidal waves at appropriate frequencies and amplitudes. Therefore, if we know the response of the ear to sinusoidal waves over a broad range of frequencies, we can evaluate the response of the ear to a wave pattern of any complexity.

An analysis of a wave shape into its sinusoidal components is shown in Fig. 12.7. The lowest frequency in the wave form is called the *fundamental*, and the higher frequencies are called *harmonics*. Figure 12.8, shows the sound pattern for a specific note played by various instruments. It is the harmonic content of the sound that differentiates one sound source from another. For a given note played by the various instruments shown in Fig. 12.8, the

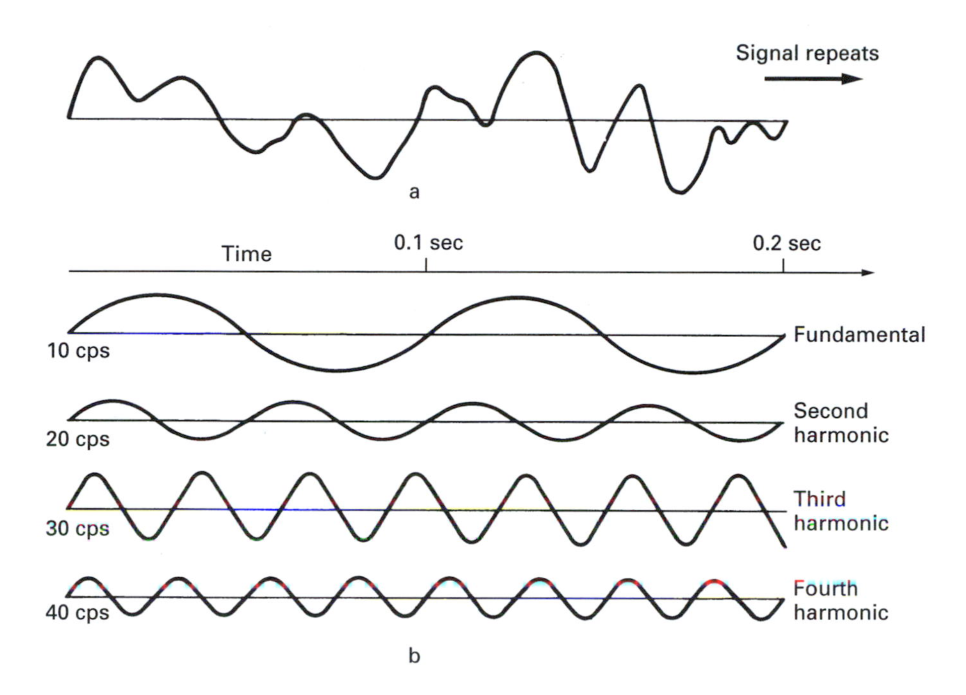

FIGURE 12.7 ▶ The analysis of a complex wave shape (a), into its sine components (b). The point-by-point addition of the fundamental frequency sine wave and the harmonic frequency sine waves yields the wave shape shown in (a).

fundamental frequency is the same but the harmonic content of the wave is different for each instrument.

12.3.2 Frequency and Pitch

The human ear is capable of detecting sound at frequencies between about 20 and 20,000 Hz. Within this frequency range, however, the response of the ear is not uniform. The ear is most sensitive to frequencies between 200 and 4000 Hz, and its response decreases toward both higher and lower frequencies. There are wide variations in the frequency response of individuals. Some people cannot hear sounds above 8000 Hz, whereas a few people can hear sounds above 20,000 Hz. Furthermore, the hearing of most people deteriorates with age.

The sensation of pitch is related to the frequency of the sound. The pitch increases with frequency. Thus, the frequency of middle C is 256 Hz, and the

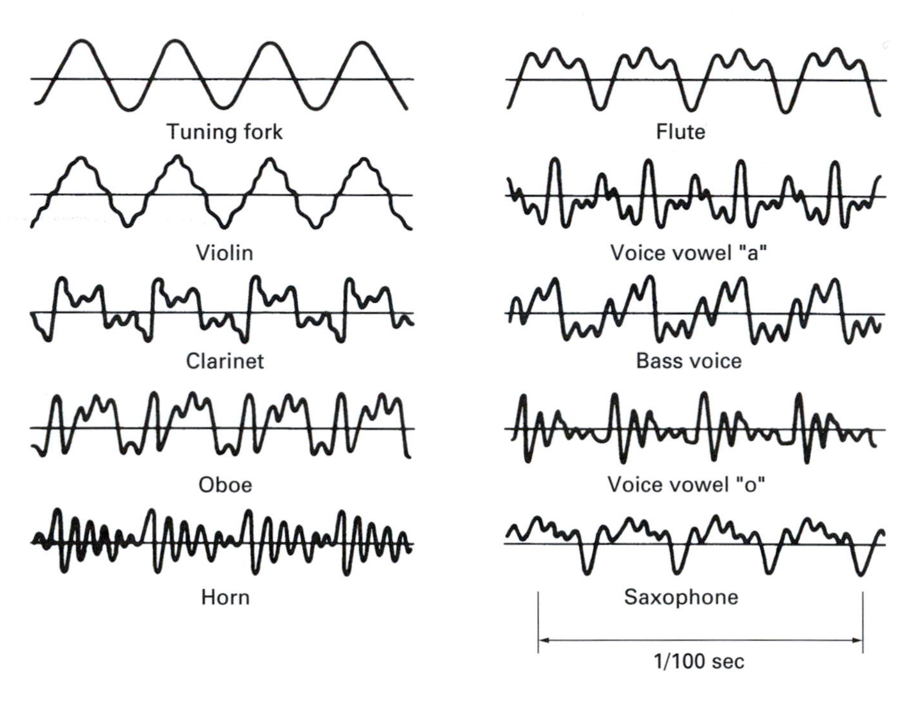

FIGURE 12.8 ▶ Wave forms of sound from different musical instruments sounding the same note.

frequency of the A above is 440 Hz. There is, however, no simple mathematical relationship between pitch and frequency.

12.3.3 Intensity and Loudness

The ear responds to an enormous range of intensities. At 3000 Hz, the lowest intensity that the human ear can detect is about 10^{-16} W/cm^2. The loudest tolerable sound has an intensity of about 10^{-4} W/cm^2. These two extremes of the intensity range are called the *threshold of hearing* and the *threshold of pain*, respectively. Sound intensities above the threshold of pain may cause permanent damage to the eardrum and the ossicles.

The ear does not respond linearly to sound intensity; that is, a sound which is a million times more powerful than another does not evoke a million times higher sensation of loudness. The response of the ear to intensity is closer to being logarithmic than linear.

Because of the nonlinear response of the ear and the large range of intensities involved in the process of hearing, it is convenient to express sound intensity on a logarithmic scale. On this scale, the sound intensity is measured

TABLE 12.1 ▶ Sound Levels Due to Various Sources (representative values)

Source of sound	Sound level (dB)	Sound level (W/cm^2)
Threshold of pain	120	10^{-4}
Riveter	90	10^{-7}
Busy street traffic	70	10^{-9}
Ordinary conversation	60	10^{-10}
Quiet automobile	50	10^{-11}
Quiet radio at home	40	10^{-12}
Average whisper	20	10^{-14}
Rustle of leaves	10	10^{-15}
Threshold of hearing	0	10^{-16}

relative to a reference level of 10^{-16} W/cm^2 (which is approximately the lowest audible sound intensity). The logarithmic intensity is measured in units of decibel (dB) and is defined as

$$\text{Logarithmic intensity} = 10 \log \frac{\text{Sound intensity in W/cm}^2}{10^{-16}\ \text{W/cm}^2} \tag{12.5}$$

Thus, for example, the logarithmic intensity of a sound wave with a power of 10^{-12} W/cm^2 is

$$\text{Logarithmic intensity} = 10 \log \frac{10^{-12}}{10^{-16}} = 40\ \text{dB}$$

Intensities of some common sounds are listed in Table 12.1.

At one time, it was believed that the ear responded logarithmically to sound intensity. Referring to Table 12.1, a logarithmic response would imply that, for example, a busy street sounds only six times louder than the rustle of leaves even though the power of the street sounds is a million times greater. Although it has been shown that the intensity response of the ear is not exactly logarithmic, the assumption of a logarithmic response still provides a useful guide for assessing the sensation of loudness produced by sounds at different intensities (see Exercises 12-1 and 12-2).

The sensitivity of the ear is remarkable. At the threshold of hearing, in the range of 2000–3000 Hz, the ear can detect a sound intensity of 10^{-16} W/cm^2. This corresponds to a pressure variation in the sound wave of only about 2.9×10^{-4} dyn/cm^2 (see Exercise 12-3). Compare this to the background atmospheric pressure, which is 1.013×10^6 dyn/cm^2. This sensitivity appears even more remarkable when we realize that the random pressure variations in air due to the thermal motion of molecules are about 0.5×10^{-4} dyn/cm^2. Thus, the sensitivity of the ear is close to the ultimate limit at which it would begin to detect the noise fluctuations in the air. The displacement of the molecules corresponding to the power at the threshold of hearing is less than the size of the molecules themselves.

The sensitivity of the ear is partly due to the mechanical construction of the ear, which amplifies the sound pressure. Most of the mechanical amplification is produced by the middle ear. The area of the eardrum is about 30 times larger than the oval window. Therefore, the pressure on the oval window is increased by the same factor (see Exercise 12-4). Furthermore, the ossicles act as a lever with a mechanical advantage of about 2. Finally, in the frequency range around 3000 Hz, there is an increase in the pressure at the eardrum due to the resonance of the ear canal. In this frequency range, the pressure is increased by another factor of 2. Thus, the total mechanical amplification of the sound pressure in the 3000-Hz range is about $2 \times 30 \times 2 = 120$. Because the intensity is proportional to pressure squared (see Eq. 12.3), the intensity at the oval window is amplified by a factor of about 14,400.

The process of hearing cannot be fully explained by the mechanical construction of the ear. The brain itself plays an important role in our perception of sound. For example, the brain can effectively filter out ambient noise and allow us to separate meaningful sounds from a relatively loud background din. (This feature of the brain permits us to have a private conversation in the midst of a loud party.) The brain can also completely suppress sounds that appear to be meaningless. Thus, we may lose awareness of a sound even though it still produces vibrations in our ear. The exact mechanism of interaction between the brain and the sensory organs is not yet fully understood.

12.4 Bats and Echoes

The human auditory organs are very highly developed; yet, there are animals that can hear even better than we can. Notable among these animals are the bats. They emit high-frequency sound waves and detect the reflected sounds (echoes) from surrounding objects. Their sense of hearing is so acute that they can obtain information from echoes which is in many ways as detailed as the information we can obtain with our sense of sight. The many different species of bats utilize echoes in various ways. The *Vespertilionidae* family of bats emit short chirps as they fly. The chirps last about 3×10^{-3} sec (3 msec) with a time interval between chirps of about 70 msec. Each chirp starts at a frequency of about 100×10^3 Hz and falls to about 30×10^3 Hz at the end. (The ears of bats, of course, respond to these high frequencies.) The silent interval between chirps allows the bat to detect the weak echo without interference from the primary chirp. Presumably the interval between the chirp and the return echo allows the bat to determine its distance from the object. It is also possible that differences in the frequency content of the chirp and the echo allow the bat to estimate the size of the object (see Exercise 12-5). With a spacing between chirps of 70 msec, an echo from an object as far as 11.5 m can be detected before the next chirp (see Exercise 12-6). As the bat comes closer to the object (such as an obstacle or an insect), both the duration of and the spacing between chirps decrease, allowing the bat to localize the object more accurately. In the final approach to the object, the duration of the chirps is only about 0.3 msec, and the spacing between them is about 5 msec.

Experiments have shown that with echo location bats can avoid wire obstacles with diameters down to about 0.1 mm, but they fail to avoid finer wires. This is in accord with our discussion of wave diffraction (see Exercise 12-7). Other animals, such as porpoises, whales, and some birds, also use echoes to locate objects, but they are not able to do so as well as bats.

12.5 Sounds Produced by Animals

Animals can make sounds in various ways. Some insects produce sounds by rubbing their wings together. The rattlesnake produces its characteristic sound by shaking its tail. In most animals, however, sound production is associated with the respiratory mechanism. In humans, the *vocal cords* are the primary source of sound. These are two reeds, shaped like lips, attached to the upper part of the trachea. During normal breathing the cords are wide open. To produce a sound the edges of the cords are brought together. Air from the lungs passes through the space between the edges and sets the cords into vibration. The frequency of the sounds is determined by the tension on the vocal cords. The fundamental frequency of the average voice is about 140 Hz for males and about 230 Hz for females. The sound produced by the vocal cords is substantially modified as it travels through the passages of the mouth and throat. The tongue also plays an important role in the final sound. Many voice sounds are produced outside the vocal cords (for example, the consonant *s*). The sounds in a whispering talk are also produced outside the vocal cords.

12.6 Clinical Uses of Sound

The most familiar clinical use of sound is in the analysis of body sounds with a stethoscope. This instrument consists of a small bell-shaped cavity attached to a hollow flexible tube. The bell is placed on the skin over the source of the body sound (such as the heart or lungs). The sound is then conducted by the pipe to the ears of the examiner who evaluates the functioning of the organ. A modified version of the stethoscope consists of two bells that are placed on different parts of the body. The sound picked up by one bell is conducted to one ear, and the sound from the other bell is conducted to the other ear. The two sounds are then compared. With this device, it is possible, for example, to listen simultaneously to the heartbeats of the fetus and of the pregnant mother.

12.7 Ultrasonic Waves

With special electronically driven crystals, it is possible to produce mechanical waves at very high frequencies, up to millions of cycles per second. These waves, which are simply the extension of sound to high frequencies, are called *ultrasonic waves*. Because of their short wavelength, ultrasonic waves can be focused onto small areas and can be imaged much as visible light (see Exercise 12-8).

Ultrasonic waves penetrate tissue and are scattered and absorbed within it. Using specialized techniques called *ultrasound imaging*, it is possible to form visible images of ultrasonic reflections and absorptions. Therefore, structures within living organisms can be examined with ultrasound, as with X-rays. Ultrasonic examinations are safer than X-rays and often can provide as much information. In some cases, such as in the examination of a fetus and the heart, ultrasonic methods can show motion, which is very useful in such displays.

The frequency of sound detected by an observer depends on the relative motion between the source and the observer. This phenomenon is called the *Doppler effect*. It can be shown (see Exercise 12-9) that if the observer is stationary and the source is in motion, the frequency of the sound f' detected by the observer is given by

$$f' = f \frac{v}{v \mp v_s} \tag{12.6}$$

where f is the frequency in the absence of motion, v is the speed of sound, and v_s is the speed of the source. The minus sign in the denominator is to be used when the source is approaching the observer, and the plus sign when the source is receding.

Using the Doppler effect, it is possible to measure motions within a body. One device for obtaining such measurements is the *ultrasonic flow meter*, which produces ultrasonic waves that are scattered by blood cells flowing in the blood vessels. The frequency of the scattered sound is altered by the Doppler effect. The velocity of blood flow is obtained by comparing the incident frequency with the frequency of the scattered ultrasound.

Within the tissue, the mechanical energy in the ultrasonic wave is converted to heat. With a sufficient amount of ultrasonic energy, it is possible to heat selected parts of a patient's body more efficiently and evenly than can be done with conventional heat lamps. This type of treatment, called *diathermy*, is used to relieve pain and promote the healing of injuries. It is actually possible to destroy tissue with very high-intensity ultrasound. Ultrasound is now routinely used to destroy kidney and gall stones (lithotripsy).

► EXERCISES ►

12-1. The intensity of a sound produced by a point source decreases as the square of the distance from the source. Consider a riveter as a point source of sound and assume that the intensities listed in Table 12.1 are measured at a distance 1 m away from the source. What is the maximum distance at which the riveter is still audible? (Neglect losses due to energy absorption in the air.)

12-2. Referring to Table 12.1, approximately how much louder does busy street traffic sound than a quiet radio?

12-3. Calculate the pressure variation corresponding to a sound intensity of 10^{-16} W/cm^2. (The density of air at 0°C and 1 atm pressure is 1.29×10^{-3} g/cm^3; for the speed of sound use the value 3.3×10^4 cm/sec.)

12-4. Explain why the relative sizes of the eardrum and the oval window result in pressure magnification in the inner ear.

12-5. Explain how a bat might use the differences in the frequency content of its chirp and echo to estimate the size of an object.

12-6. With a 70-msec space between chirps, what is the farthest distance at which a bat can detect an object?

12-7. In terms of diffraction theory, discuss the limitations on the size of the object that a bat can detect with its echo location.

12-8. Estimate the lower limit on the size of objects that can be detected with ultrasound at a frequency of 2×10^6 Hz.

12-9. With the help of a basic physics textbook, explain the Doppler effect and derive Eq. 12.6.

Chapter 13

Electricity

The word *electricity* usually evokes the image of a man-made technology because we usually associate electricity with devices such as amplifiers, televisions, and computers. This technology has certainly played an important role in our understanding of living systems, as it has provided the major tools for the study of life processes. However, many life processes themselves involve electrical phenomena. The nervous system of animals and the control of muscle movement, for example, are both governed by electrical interactions. Even plants rely on electrical forces for some of their functions. In this chapter, we will describe some of the electrical phenomena in living organisms, and in Chapter 14 we will discuss the applications of electrical technology in biology and medicine. A brief review of electricity in Appendix B summarizes the concepts, definitions, and equations used in the text.

13.1 The Nervous System

The most remarkable use of electrical phenomena in living organisms is found in the nervous system of animals. Specialized cells called *neurons* form a complex network within the body which receives, processes, and transmits information from one part of the body to another. The center of this network is located in the brain, which has the ability to store and analyze information. Based on this information, the nervous system controls various parts of the body. The nervous system is very complex. The human nervous system, for example, consists of about 10^{10} interconnected neurons. It is, therefore, not surprising that, although the nervous system has been studied for more than a hundred years, its functioning as a whole is still poorly understood. It is not known how information is stored and processed by the nervous system; nor is it known how the neurons grow into patterns specific to their functions. Yet some aspects of the nervous system are now well known. Specifically, during the past 40 years, the method of signal propagation through the nervous system has been firmly established. The messages are electrical pulses transmitted by the neurons. When a neuron receives an appropriate stimulus, it produces electrical pulses that are propagated along its cablelike structure. The pulses are constant in magnitude and duration, independent of the intensity of the stimulus. The strength of the stimulus is conveyed by the number of pulses produced. When the pulses reach the end of the "cable," they activate other neurons or muscle cells.

13.1.1 The Neuron

The neurons, which are the basic units of the nervous system, can be divided into three classes: *sensory* neurons, *motor* neurons, and *interneurons*. The sensory neurons receive stimuli from sensory organs that monitor the external and internal environment of the body. Depending on their specialized functions, the sensory neurons convey messages about factors such as heat, light, pressure, muscle tension, and odor to higher centers in the nervous system for processing. The motor neurons carry messages that control the muscle cells. These messages are based on information provided by the sensory neurons and by the central nervous system located in the brain. The interneurons transmit information between neurons.

Each neuron consists of a cell body to which are attached input ends called *dendrites* and a long tail called the *axon* which propagates the signal away from the cell (see Fig. 13.1). The far end of the axon branches into nerve endings that transmit the signal across small gaps to other neurons or muscle cells. A simple sensory-motor neuron circuit is shown in Fig. 13.2. A stimulus from a muscle produces nerve impulses that travel to the spine. Here the signal is transmitted to a motor neuron, which in turn sends impulses to control the muscle. Such simple circuits are often associated with reflex actions. Most nervous connections are far more complex.

The axon, which is an extension of the neuron cell, conducts the electrical impulses away from the cell body. Some axons are long indeed—in people, for example, the axons connecting the spine with the fingers and toes are more than a meter in length. Some of the axons are covered with a segmented sheath of fatty material called *myelin*. The segments are about 2 mm long, separated by gaps called the *Nodes of Ranvier*. We will show later that the myelin sheath increases the speed of pulse propagation along the axon.

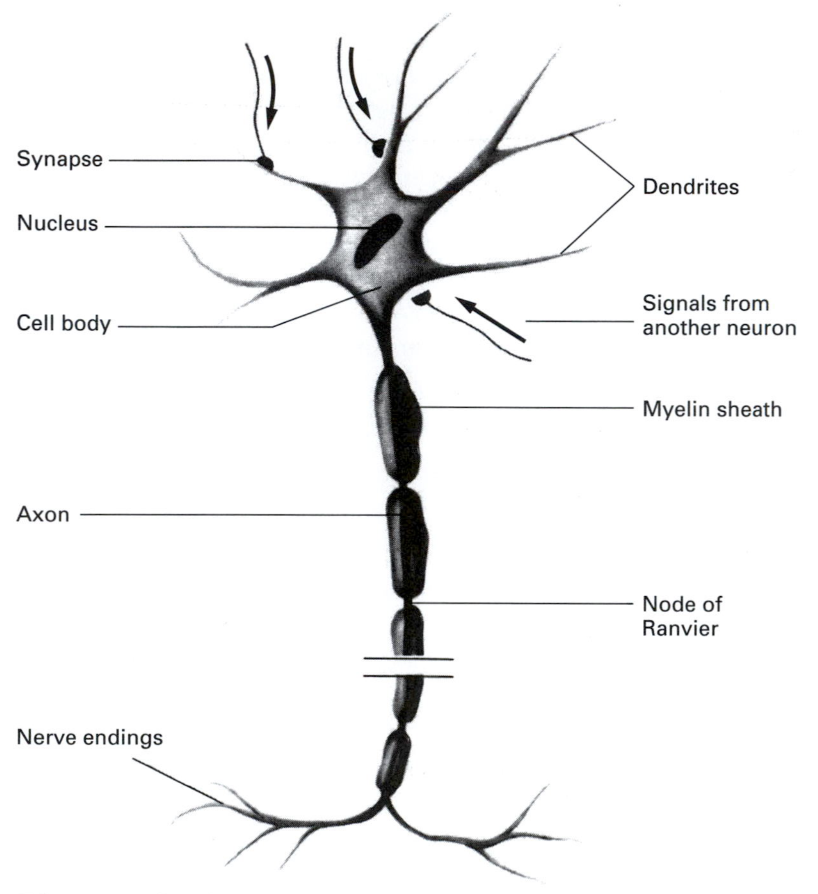

FIGURE 13.1 ▶ A neuron.

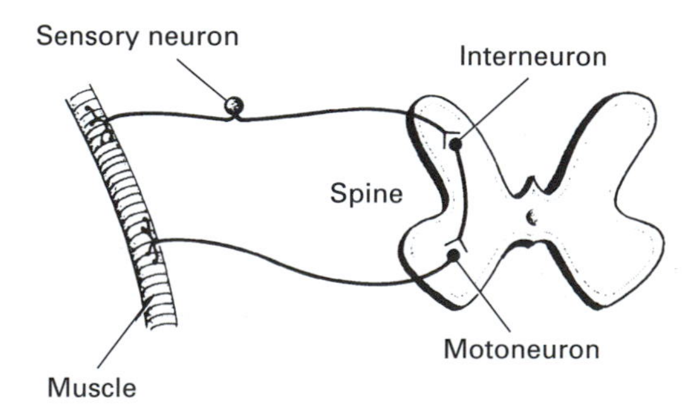

FIGURE 13.2 ▶ A simple neural circuit.

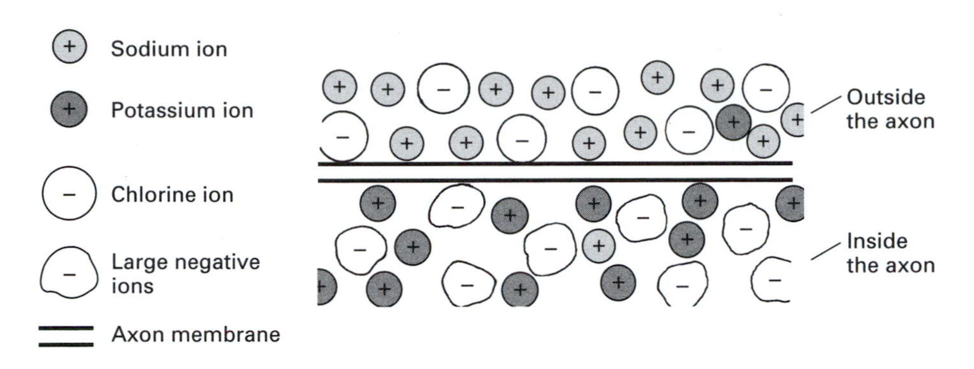

FIGURE 13.3 ▶ The axon membrane and surroundings.

Although each axon propagates its own signal independently, many axons often share a common path within the body. These axons are usually grouped into nerve bundles.

The ability of the neuron to transmit messages is due to the special electrical characteristics of the axon. Most of the data about the electrical and chemical properties of the axon is obtained by inserting small needlelike probes into the axon. With such probes it is possible to measure currents flowing in the axon and to sample its chemical composition. Such experiments are usually difficult as the diameter of most axons is very small. Even the largest axons in the human nervous system have a diameter of only about 20 μm (20×10^{-4} cm). The squid, however, has a giant axon with a diameter of about 500 μm (0.5 mm), which is large enough for the convenient insertion of probes. Much of the information about signal transmission in the nervous system has been obtained from experiments with the squid axon.

13.1.2 Electrical Potentials in the Axon

In the aqueous environment of the body, salt and various other molecules dissociate into positive and negative ions. As a result, body fluids are relatively good conductors of electricity. Still, these fluids are not nearly as conductive as metals; their resistivity is about 100 million times greater than that of copper, for example.

The inside of the axon is filled with an ionic fluid that is separated from the surrounding body fluid by a thin membrane (Fig. 13.3). The axon membrane, which is only about 50–100 Å thick, is a relatively good but not perfect electrical insulator. Therefore, some current can leak through it.

The electrical resistivities of the internal and the external fluids are about the same, but their chemical compositions are substantially different. The external fluid is similar to sea water. Its ionic solutes are mostly positive sodium ions and negative chlorine ions. Inside the axon, the positive ions are mostly potassium ions, and the negative ions are mostly large negatively charged organic molecules.

Because there is a large concentration of sodium ions outside the axon and a large concentration of potassium ions inside the axon, we may ask why the concentrations are not equalized by diffusion. In other words, why don't the sodium ions leak into the axon and the potassium ions leak out of it? The answer lies in the properties of the axon membrane.

In the resting condition, when the axon is not conducting an electrical pulse, the axon membrane is highly permeable to potassium and only slightly permeable to sodium ions. The membrane is impermeable to the large organic ions. Thus, while sodium ions cannot easily leak in, potassium ions can certainly leak out of the axon. However, as the potassium ions leak out of the axon, they leave behind the large negative ions, which cannot follow them through the membrane. As a result, a negative potential is produced inside the axon with respect to the outside. This negative potential, which has been measured to be about 70 mV, holds back the outflow of potassium so that in equilibrium the concentration of ions is as we have stated. Some sodium ions do in fact leak into the axon, but they are continuously removed by a metabolic mechanism called the *sodium pump*. This pumping process, which is not yet fully understood, transports sodium ions out of the cell and brings in an equal number of potassium ions.

13.1.3 Action Potential

The description of the axon that we have so far given applies to other types of cells as well. Most cells contain an excess concentration of potassium ions and are at a negative potential with respect to their surroundings. The special property of the neuron is its ability to conduct electrical impulses.

Physiologists have studied the properties of nerve impulses by inserting a probe into the axon and measuring the changes in the axon voltage with respect to the surrounding fluid. The nerve impulse is elicited by some stimulus on the neuron or the axon itself. The stimulus may be an injected chemical, mechanical pressure, or an applied voltage. In most experiments the stimulus is an externally applied voltage, as shown in Fig. 13.4.

A nerve impulse is produced only if the stimulus exceeds a certain threshold value. When this value is exceeded, an impulse is generated at the point of stimulation and propagates down the axon. Such a propagating impulse is called an *action potential*. An action potential as a function of time at one point on the axon is shown in Fig. 13.5. The scales of time and voltage are typical of most neurons. The arrival of the impulse is marked by a sudden

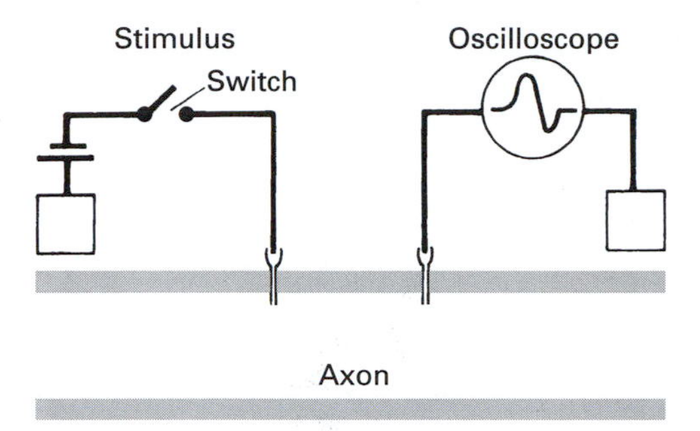

FIGURE 13.4 ▶ Measuring the electrical response of the axon.

rise of the potential inside the axon from its negative resting value to about +30 mV. The potential then rapidly decreases to about −90 mV and returns more slowly back to the initial resting state. The whole pulse passes a given point in a few milliseconds. The speed of pulse propagation depends on the type of axon. Fast-acting axons propagate the pulse at speeds up to 100 m/sec. The mechanism for the action potential is discussed in a following section.

Impulses produced by a given neuron are always of the same size and propagate down the axon without attenuation. The nerve impulses are produced at a rate proportional to the intensity of the stimulus. There is, however,

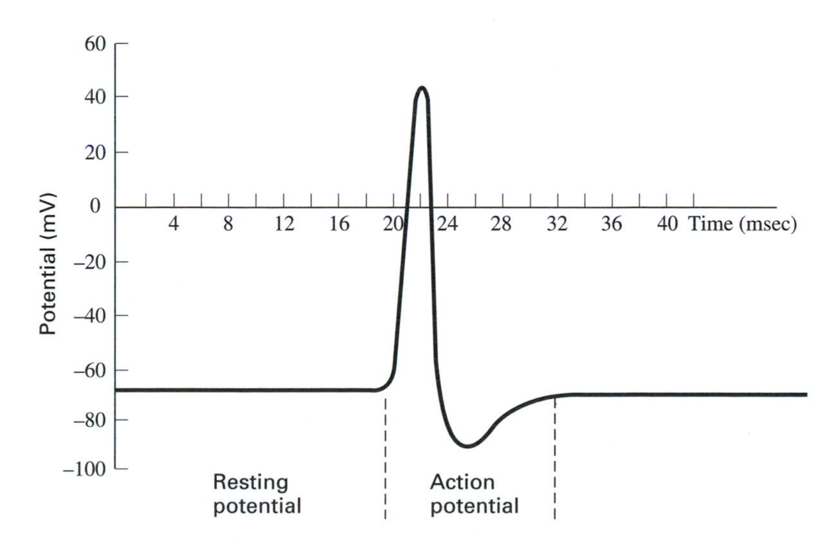

FIGURE 13.5 ▶ The action potential.

an upper limit to the frequency of impulses because a new impulse cannot begin before the previous one is completed.

13.1.4 Axon as an Electric Cable

In the analysis of the electrical properties of the axon, we will use some of the techniques of electrical engineering. This treatment is more complex than the methods used in the other sections of the text. The added complexity, however, is necessary for the quantitative understanding of the nervous system.

Although the axon is often compared to an electrical cable, there are profound differences between the two. Still, it is possible to gain some insight into the functioning of the axon by analyzing it as an insulated electric cable submerged in a conducting fluid. In such an analysis, we must take into account the resistance of the fluids both inside and outside the axon and the electrical properties of the axon membrane. Because the membrane is a leaky insulator, it is characterized by both capacitance and resistance. Thus, we need four electrical parameters to specify the cable properties of the axon.

The capacitance and the resistance of the axon are distributed continuously along the length of the cable. It is, therefore, not possible to represent the whole axon (or any other cable) with only four circuit components. We must consider the axon as a series of very small electrical-circuit sections joined together. When a potential difference is set up between the inside and the outside of the axon, four currents can be identified: the current outside the axon, the current inside the axon, the current through the resistive component of the membrane, and the current through the capacitive component of the membrane (see Fig. 13.6). The electrical circuit representing a small axon section of length Δx is shown in Fig. 13.6. In this small section, the resistances of the outside and of the inside fluids are R_o and R_i, respectively. The membrane capacitance and resistance are shown as C_m and R_m. The whole axon is just a long series of these subunits joined together. This is shown in Fig. 13.7. Sample values of the circuit parameters for both a myelinated and a nonmyelinated axon of radius 5×10^{-6} m are listed in Table 13.1. (These values were obtained from [13.4].) Note that the values in Table 13.1 are quoted for a 1-m length of the axon. The unit *mho* for the conductivity of the axon membrane is defined in Appendix B.

An examination of the axon performance shows immediately that the circuit in Fig. 13.7 does not explain some of the most striking features of the axon. An electrical signal along such a circuit propagates at nearly the speed of light (3×10^8 m/sec), whereas a pulse along an axon travels at a speed that is at most about 100 m/sec. Furthermore, as we will show, the circuit in Fig. 13.7 dissipates an electrical signal very quickly; yet we know that action potentials propagate along the axon without any attenuation. Therefore, we

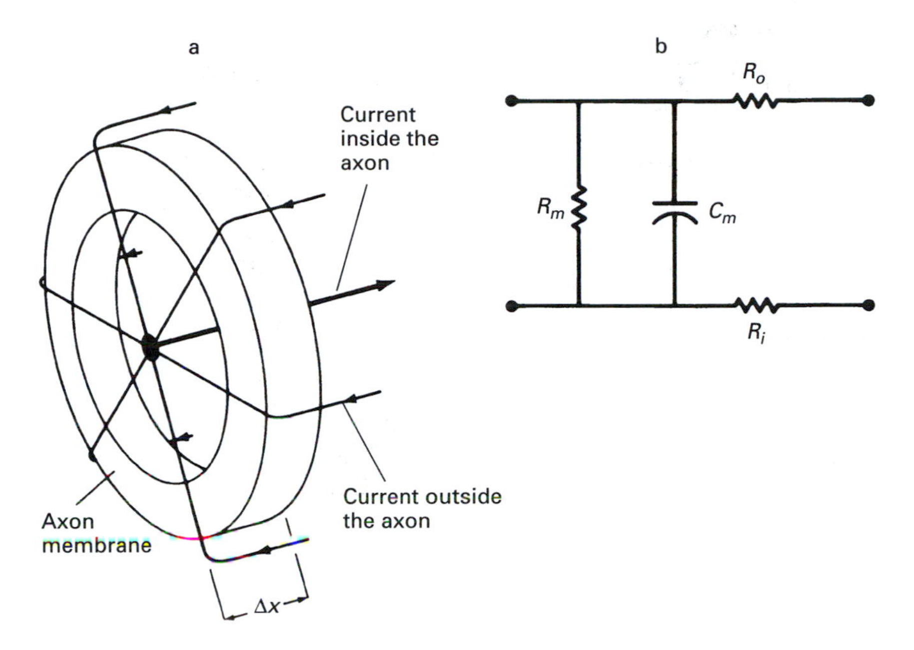

FIGURE 13.6 ▶ (a) Currents flowing through a small section of the axon. (b) Electrical circuit representing a small section of the axon.

must conclude that an electrical signal along the axon does not propagate by a simple passive process.

13.1.5 Propagation of the Action Potential

After many years of research the propagation of an impulse along the axon is now reasonably well understood. (See Fig. 13.8.) When the magnitude of the voltage across a portion of the membrane is reduced below a threshold value, the permeability of the axon membrane to sodium ions increases rapidly. As a result, sodium ions rush into the axon, cancel out the local negative charges,

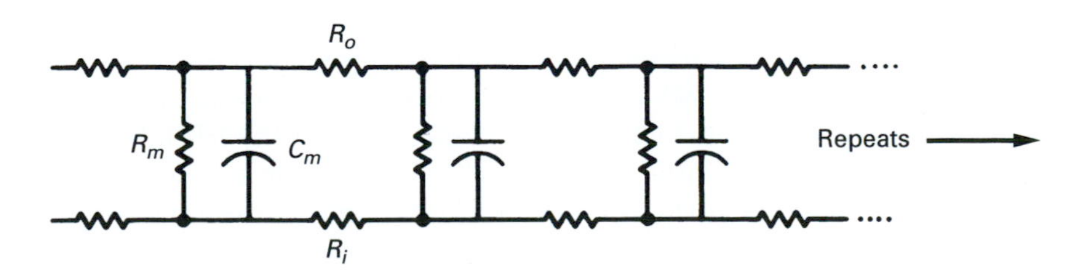

FIGURE 13.7 ▶ The axon represented as an electrical cable.

TABLE 13.1 ► Properties of Sample Axons

Property	Nonmyelinated axon	Myelinated axon
Axon radius	5×10^{-6} m	5×10^{-6} m
Resistance per unit length of fluid both inside and outside axon (r)	$6.37 \times 10^9 \Omega/\text{m}$	6.37×10^9 Ω/m
Conductivity per unit length of axon membrane (g_m)	1.25×10^{-4} mho/m	3×10^{-7} mho/m
Capacitance per unit length of axon (c)	3×10^{-7} F/m	8×10^{-10} F/m

and, in fact, drive the potential inside the axon to a positive value. This process produces the initial sharp rise of the action potential pulse. The sharp positive spike in one portion of the axon increases the permeability to the sodium immediately ahead of it which in turn produces a spike in that region. In this way the disturbance is sequentially propagated down the axon, much as a flame is propagated down a fuse.

The axon, unlike a fuse, renews itself. At the peak of the action potential, the axon membrane closes its gates to sodium and opens them wide to potassium ions. The potassium ions now rush out, and, as a result, the axon potential drops to a negative value slightly below the resting potential. After a few milliseconds, the axon potential returns to its resting state and that portion of the axon is ready to receive another pulse.

The number of ions that flow in and out of the axon during the pulse is so small that the ion densities in the axon are not changed appreciably. The cumulative effect of many pulses is balanced by metabolic pumps that keep the ion concentrations at the appropriate levels. Using Eq. B.5 in Appendix B, we can estimate the number of sodium ions that enter the axon during the rising phase of the action potential. The initial inrush of sodium ions changes the amount of electrical charge inside the axon. We can express this change in charge ΔQ in terms of the change in the voltage ΔV across the membrane capacitor C, that is,

$$\Delta Q = C \Delta V \tag{13.1}$$

In the resting state, the axon voltage is -70 mV. During the pulse, the voltage changes to about $+30$ mV, resulting in a net voltage change across the membrane of 100 mV. Therefore, ΔV to be used in Eq. 13.1 is 100 mV.

The calculations outlined in Exercise 13-1 show that, in the case of the nonmyelinated axon described in Table 13.1, during each pulse 1.87×10^{11} sodium ions enter per meter of axon length. The same number of potassium ions leaves during the following part of the action potential. (Measurements show that actually the ion flow is about three times higher than our simple

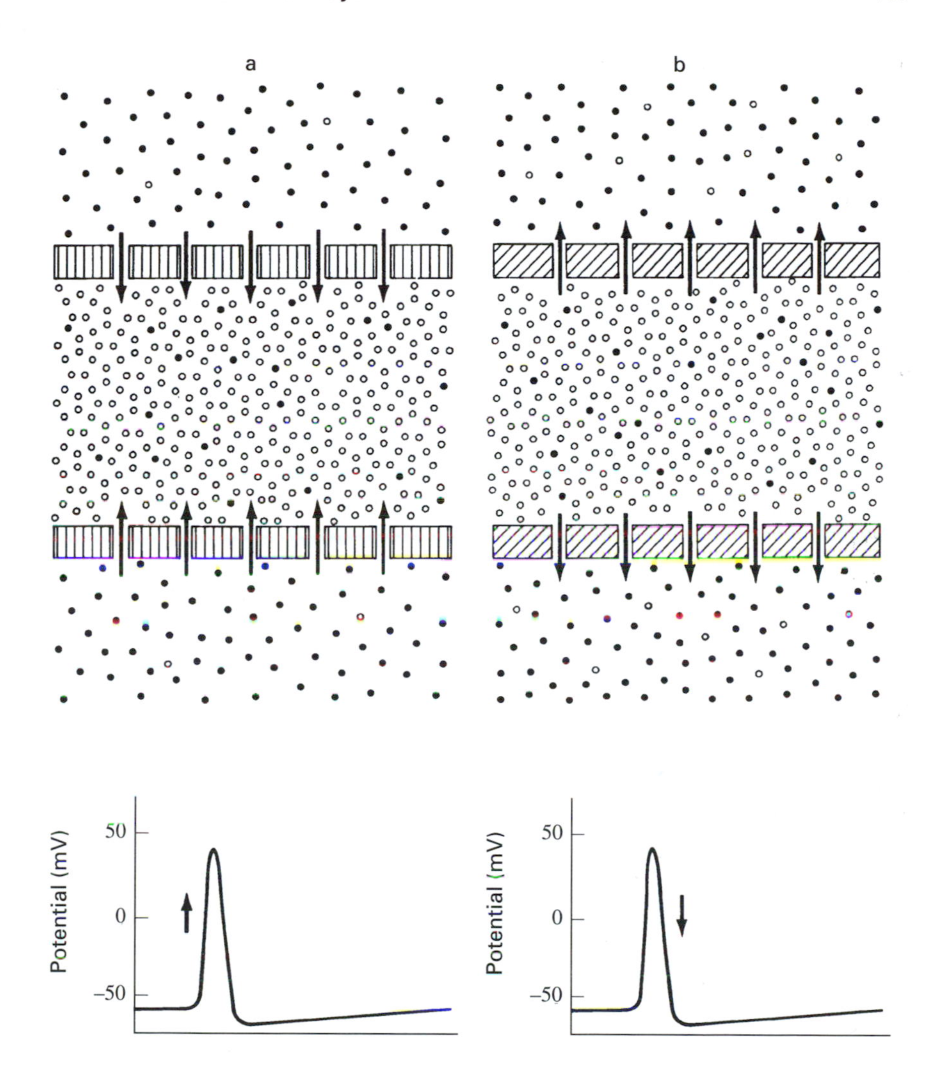

FIGURE 13.8 ▶ The action potential. (a) The action potential begins with the axon membrane becoming highly permeable to sodium ions (closed circles) which enter the axon making it positive. (b) The sodium gates then close and potassium ions (open circles) leave the axon making the interior negative again.

estimate.) The exercise also shows that, in the resting state, the number of sodium ions inside a meter length of the axon is about 7×10^{14} and the number of potassium ions is 7×10^{15}. Thus, the inflow and the outflow of ions during the action potential is negligibly small compared to the equilibrium density.

Another simple calculation using Eq. B.6 yields an estimate of the minimum energy required to propagate the impulse along the axon. During the propagation of one pulse, the whole axon capacitance is successively discharged and then must be recharged again. The energy required to recharge a meter length of the nonmyelinated axon is

$$E = \frac{1}{2}C(\Delta V)^2 = \frac{1}{2} \times 3 \times 10^{-7} \times (0.1)^2 = 1.5 \times 10^{-9}\text{J/m} \qquad (13.2)$$

where C is the capacitance per meter of the axon. Because the duration of each pulse is about 10^{-2} sec, and an axon can propagate at most 100 pulses/sec, even at peak operation the axon requires only 1.5×10^{-7} W/m to recharge its capacitance.

13.1.6 An Analysis of the Axon Circuit

The circuit in Fig. 13.7 does not contain the pulse-conducting mechanism of the axon. It is possible to incorporate this mechanism into the circuit by connecting small signal generators along the circuit. However, the analysis of such a complex circuit is outside the scope of this text. Even the circuit in Fig. 13.7 cannot be fully analyzed without calculus. We will simplify this circuit by neglecting the capacitance of the axon membrane. The circuit is then as shown in Fig. 13.9a. This representation is valid when the capacitors are fully charged so that the capacitive current is zero. With this model, we will be able to calculate the voltage attenuation along the cable when a steady voltage is applied at one end. The simplified model, however, cannot make predictions about the time-dependent behavior of the axon.

The problem then is to calculate the voltage $V(x)$ at point x when a voltage V_a is applied at point x_0 (see Fig. 13.9a). The approach is to calculate first the voltage drop across a small incremental cable section of length Δx cut by lines a and b (see [13.5, 13.6]). We assume that the cable is infinite in length and that the total cable resistance to the right of line b is R_T. Thus, the whole cable to the right of line b is replaced by R_T as shown in Fig. 13.9b. Because the cable is infinite, the resistance to the right of any vertical cut equivalent to line b is also R_T. Specifically, the resistance to the right of line a is R_T. We can, therefore, calculate R_T by equating the resistance to the right of line a in Fig. 13.9b to R_T, that is,

$$R_T = R_o + R_i + \frac{R_T R_m}{R_T + R_m} \qquad (13.3)$$

Measurements show that the resistivities inside and outside the axon are about the same. Therefore, $R_i = R_o = R$ and Eq. 13.3 simplifies to

$$R_T = 2R + \frac{R_T R_m}{R_T + R_m} \qquad (13.4)$$

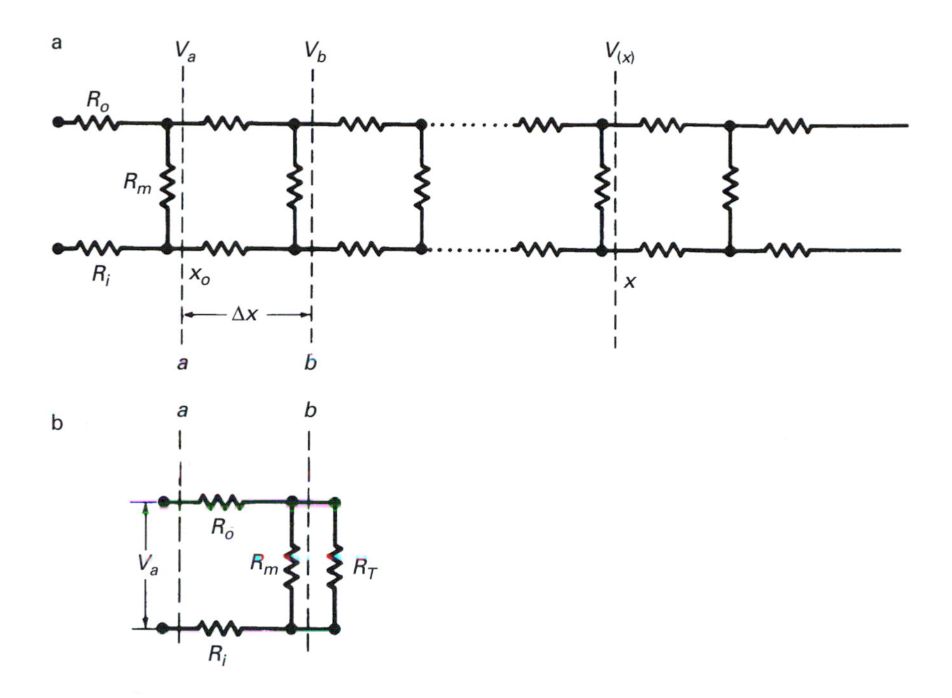

FIGURE 13.9 ▶ (a) Approximation to the circuit in Fig. 13.7 with the capacitances neglected. (b) The resistances to the right of line b replaced by the equivalent resistor R_T.

The solution of Eq. 13.4 (see Exercise 13-2) yields

$$R_T = R + \left[R^2 + 2R\ R_m\right]^{1/2} \tag{13.5}$$

A simple circuit analysis (see Exercise 13-3) shows that

$$V_b = \frac{V_a}{1 + \left[\frac{(2R)(R_T + R_m)}{R_T R_m}\right]} = \frac{V_a}{1 + \beta} \tag{13.6}$$

where β is the quantity in the square brackets.

We can calculate β from the measured parameters shown in Table 13.1. The resistances R and R_m are the values for a small axon section of length Δx. Therefore,

$$R = r\Delta x \quad \text{and} \quad \frac{1}{R_m} = g_m \Delta x \qquad \text{or} \qquad R_m = \frac{1}{g_m \Delta x}$$

From Eq. 13.5, it can be shown (see Exercise 13-4) that if Δx is very small then

$$R_T = \left(\frac{2r}{g_m}\right)^{1/2} \tag{13.7}$$

and

$$\beta = (2rg_m)^{1/2}\,\Delta x = \frac{\Delta x}{\lambda} \tag{13.8}$$

where

$$\lambda = \left(\frac{1}{2rg_m}\right)^{1/2} \tag{13.9}$$

Now returning to Eq. 13.6, since Δx is vanishingly small, β is also very small. Therefore, the term $1/(1+\beta)$ is approximately equal to $1-\beta$ (see Exercise 13-5). Consequently, the voltage V_b at b, a distance Δx away from a, is

$$V_b = V_a\left[1 - \frac{\Delta x}{\lambda}\right] \tag{13.10}$$

To obtain the voltage at a distance x away from line a, we divide this distance into increments of Δx such that $n\Delta x = x$. We can then apply Eq. 13.10 successively down the cable and obtain the voltage at x (see Exercise 13-6) as

$$V(x) = V_a\left[1 - \frac{\Delta x}{\lambda}\right]^n \tag{13.11}$$

It can be shown that, for small Δx and large n, Eq. 13.11 can be written as (see Exercise 13-7)

$$V(x) = V_a e^{-x/\lambda} \tag{13.12}$$

Equation 13.12 states that if a steady voltage V_a is applied across one point in the axon membrane, the voltage decreases exponentially down the axon. From Table 13.1, for a nonmyelinated axon λ is about 0.8 mm. Therefore, at a distance 0.8 mm from the point of application, the voltage decreases to 37% of its value at the point of application.

Myelinated axons, because of their outer sheath, have a much smaller membrane conductance than axons without myelin. As a result, the value of λ is larger. Using the values given in Table 13.1, we can show that λ is 16 mm for a myelinated axon. This result helps to explain the faster pulse conduction along myelinated axons. As we mentioned earlier, the myelin sheath is in 2-mm-long segments. The action potential is generated only at the nodes between the segments. The pulse propagates through the myelinated segments as a fast conventional electrical signal. Because λ is 16 mm, the pulse decreases by only 13% as it traverses one segment, and it is still sufficiently intense to generate an action potential at the next node.

13.1.7 Synaptic Transmission

So far we have considered the propagation of an electrical impulse down the axon. Now we shall briefly describe how the pulse is transmitted from the axon to other neurons or muscle cells.

At the far end, the axon branches into nerve endings which extend to the cells that are to be activated. Through these nerve endings the axon transmits signals, usually to a number of cells. In some cases the action potential is transmitted from the nerve endings to the cells by electrical conduction. In the vertebrate nervous system, however, the signal is usually transmitted by a chemical substance. The nerve endings are actually not in contact with the cells. There is a gap, about a nanometer wide (1 nm $= 10^{-9}$ m $= 10^{-7}$ cm) between the nerve ending and the cell body. These regions of interaction between the nerve ending and the target cell are called *synapses* (see Fig. 13.10). When the impulse reaches the synapse, a chemical substance is released at the nerve ending which quickly diffuses across the gap and stimulates the adjacent cell. The chemical is released in bundles of discrete size.

Usually a neuron is in synaptic contact with many sources. Often a number of synapses must be activated simultaneously to start the action potential in the target cell. The action potential produced by a neuron is always of the same magnitude. The neuron operates in an all-or-none mode: It either produces an action potential of the standard size or does not fire at all. In some cases the chemicals released at the synapse do not stimulate the cell but inhibit its response to impulses arriving along a different channel. Presumably, these types of interactions permit decisions to be made on a cellular level. The details of these processes are not yet fully understood.

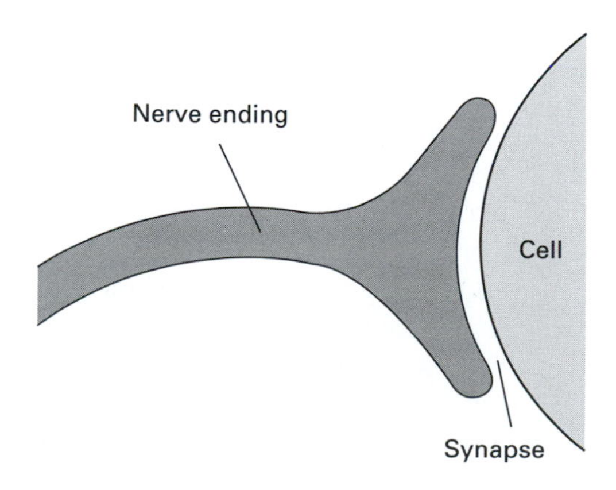

FIGURE 13.10 ► Synapse.

13.1.8 Action Potentials in Muscles

Muscle fibers produce and propagate electrical impulses in much the same way as neurons. The action potential in the muscle fiber is initiated by the impulses arriving from motor neurons. This stimulation causes a reduction of the potential across the fiber membrane which initiates the ionic process involved in the pulse propagation. The shape of the action potential is the same as in the neuron except that its duration is usually longer. In skeletal muscles, the action potential lasts about 20 msec, whereas in heart muscles it may last a quarter of a second.

After the action potential passes through the muscle fiber, the fiber contracts. In Chapter 5, we briefly discussed some aspects of muscle contraction. The details of this process are not yet fully understood.

Within the skeletal muscle fibers, mechanoreceptor organs called *muscle spindles* continuously transmit information on the state of muscle contraction. This information is relayed via neurons for processing and further action. In this way, the movement of muscles is continuously under control.

It is possible to stimulate muscle fibers by an external application of an electric current. This effect was first observed in 1780 by Luigi Galvani who noted that a frog's leg twitched when an electric current passed through it. (Galvani's initial interpretation of this effect was wrong.) External muscle stimulation is a useful clinical technique for maintaining muscle tone in cases of temporary muscle paralysis resulting from nerve disorders.

13.1.9 Surface Potentials

The voltages and currents associated with the electrical activities in neurons, muscle fibers, and other cells extend to regions outside the cells. As an example, consider the propagation of the action potential along the axon. As the voltage at one point on the axon drops suddenly, a voltage difference is produced between this point and the adjacent regions. Consequently, current flows both inside and outside the axon (see Fig. 13.11). As a result, a voltage drop develops along the outer surface length of the axon.

Experiments are sometimes performed on a whole nerve consisting of many axons. As shown in Fig. 13.12, two electrodes are placed along the nerve bundle, and the voltage between them is recorded. This measured voltage is the sum of the surface potentials produced by the individual axons and yields some information about the collective behavior of the axons.

Electric fields associated with the activities in cells extend all the way to the surface of the animal body. Thus, along the surface of the skin, we can measure electric potentials representing the collective cell activities associated with certain processes in the body. Based on this effect, clinical techniques have been developed to obtain, from the skin surface, information about the

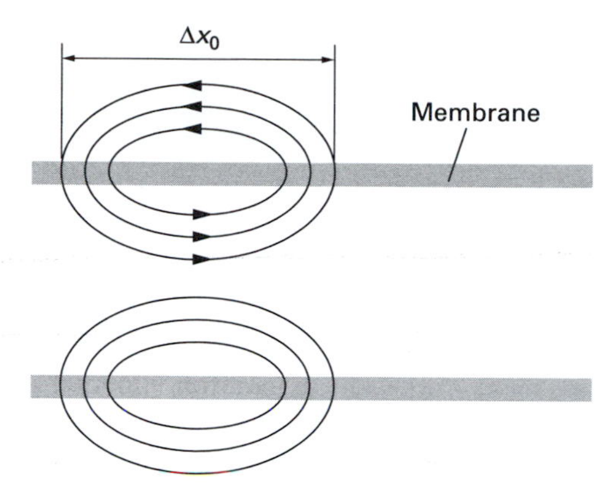

FIGURE 13.11 ▶ The action potential produces currents both inside and outside the axon.

activities of the heart (*electrocardiography*) and the brain (*electroencephalography*). The measurement of these surface signals will be discussed in Chapter 14.

Surface signals are associated with many other activities, such as movement of the eye, contractions of the gastrointestinal tract, and movement of muscles. Using a technique called *electromyography* (EMG), measurement of action potentials and their speed of propagation along muscles can provide information about muscle and nerve disorders. (See [13-5].) Surface potentials associated with metabolic activities have also been observed in plants and bones, as discussed in the following sections.

13.2 Electricity in Plants

The type of propagating electrical impulses we have discussed in connection with neurons and muscle fibers have also been found in certain plant cells. The

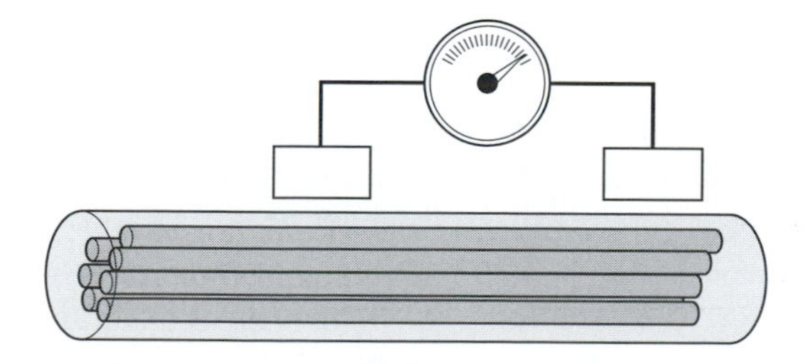

FIGURE 13.12 ▶ Surface potential along a nerve bundle.

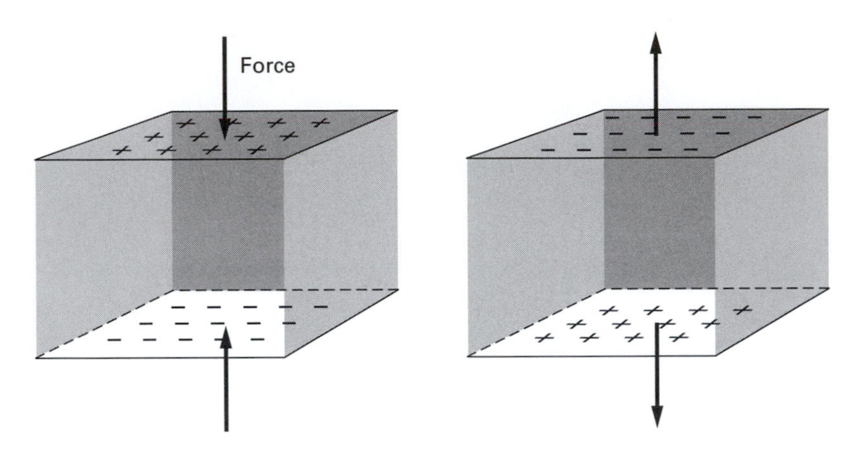

FIGURE 13.13 ▶ The piezoelectric effect.

shape of the action potential is the same in both cases, but the duration of the action potential in plant cells is a thousand times longer, lasting about 10 sec. The speed of propagation of these plant action potentials is also rather slow, only a few centimeters per second. In plant cells, as in neurons, the action potential is elicited by various types of electrical, chemical, or mechanical stimulation. However, the initial rise in the plant cell potential is produced by an inflow of calcium ions rather than sodium ions.

The role of action potentials in plants is not yet known. It is possible that they coordinate the growth and the metabolic processes of the plant and perhaps control the long-term movements exhibited by some plants.

13.3 Electricity in the Bone

When certain types of crystals are mechanically deformed, the charges in them are displaced; as a result, they develop voltages along the surface. This phenomenon is called the *piezoelectric effect* (Fig. 13.13). Bone is composed of a crystalline material that exhibits the piezoelectric effect. It has been suggested that these piezoelectric voltages play a role in the formation and nourishment of the bone.

The body has mechanisms for both building and destroying bone. New bone is formed by cells called *osteoblasts* and is dissolved by cells called *osteoclasts*. It has been known for some time that a living bone will adapt its structure to a long-term mechanical load. For example, if a compressive force bends a bone, after a time the bone will assume a new shape. Some portions of the bone gain substance and others lose it in such a way as to strengthen the bone in its new position. It has been suggested that the appropriate addition

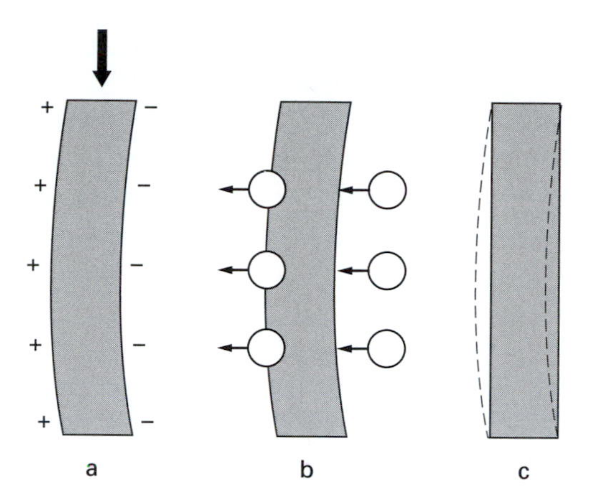

FIGURE 13.14 ▶ (a) Charges are produced along the bone as a result of deformation. (b) Guided by the charge distribution, substance is added and removed from the bone. (c) Reconstructed bone.

and removal of bone tissue is guided by the piezoelectric potentials produced by the deformation (see Fig. 13.14).

Piezoelectricity in the bone may have yet another function. All tissue including bone has to be nourished by fluids. The nourishing fluids move into the bone through very narrow canals. Without a pumping mechanism, the flow of this fluid would be too slow to provide the bone with the necessary nutrients. It has been suggested that the piezoelectric voltages produced by forces due to normal body movement act on the ions in the nutrient fluid and pump it in and out of the bone.

13.4 Electric Fish

Most animals do not possess sensory organs that are specifically designed to detect external electric fields, but sharks and rays are exceptions. These fish have small organs along their skin which are remarkably sensitive to electric fields in water. A shark responds to an electric field as small as 1 μV/m, which is in the range of fields found along the skin of animals. (A flashlight battery with terminals separated by 1500 km produces a field of this strength.) The shark uses these electrical organs to locate animals buried in sand and perhaps even to communicate with other sharks. Sharks are also known to bite boat propellers, probably in response to the electric field generated in the proximity of the metal.

An equally remarkable use of electricity is found in the electric eel, which can generate along its skin electric pulses up to 500 V with currents reaching

80 mA. The eel uses this ability as a weapon. When it comes in contact with its prey, the high-voltage pulse passes through the victim and stuns it.

The electric organ of the eel consists of specialized muscle fibers that are connected together electrically. The high voltage is produced by a series interconnection of many cells, and the large current is obtained by connecting the series chains in parallel (see Exercise 13-8). A number of other fish possess similar electric organs.

► EXERCISES ►

13-1. (a) Using Eq. 13.1 and the data in Table 13.1, calculate the number of ions entering the axon during the action potential, per meter of nonmyelinated axon length. (The charge on the ion is 1.6×10^{-19} coulomb.) (b) During the resting state of the axon, typical concentrations of sodium and potassium ions inside the axon are 15 and 150 millimole/liter, respectively. From the data in Table 13.1, calculate the number of ions per meter length of the axon.

$$1 \text{ mole/liter} = 6.02 \times 10^{20} \frac{\text{particles (ions, atoms, etc.)}}{\text{cm}^3}$$

13-2. From Eq. 13.4, obtain a solution for R_T. (Remember that R_T must be positive.)

13-3. Verify Eq. 13.6.

13-4. Show that when Δx is very small, R_T is given by Eq. 13.7.

13-5. Show that if β is small, $1/(1+\beta) \approx 1 - \beta$. (Refer to tables of series expansion.)

13-6. Verify Eq. 13.11.

13-7. Using the binomial theorem, show that Eq. 13.11 can be written as

$$V(x) = V_a \left[1 - \frac{n\Delta x}{\lambda} + \frac{n(n-1)}{2!} \left(\frac{\Delta x}{\lambda} \right)^2 - \frac{n(n-1)(n-2)}{3!} \left(\frac{\Delta x}{\lambda} \right)^3 + \cdots \right]$$

Since Δx is vanishingly small, n must be very large. Show that the above equation approaches the expansion for an exponential function. (Refer to tables of series expansion.)

13-8. (a) From the data provided in the text, estimate the number of cells that must be connected in series to provide the 500 V observed at the skin of the electric eel. (b) Estimate the number of chains that must be connected in parallel to provide the observed currents.

Assume that the size of the cell is 10^{-5} m, the pulse produced by a single cell is 0.1 V, and the duration of pulse is 10^{-2} sec. Use the data in the text and in Exercise 13-1 to estimate the current flowing into a single cell during the action potential.

Chapter 14

Electrical Technology

Electrical technology was developed by applying some of the basic principles of physics to problems in communications and industry. Although this technology was directed primarily toward industrial and military applications, it has made great contributions to the life sciences. Electrical technology has provided tools for the observation of biological phenomena that would have been otherwise inaccessible. It yielded most of the modern clinical and diagnostic equipment used in medicine. Even the techniques developed for the analysis of electrical devices have been useful in the study of living systems. In this chapter, we will describe some of the applications of electrical technology in these areas.

14.1 Electrical Technology in Biological Research

Our understanding of the world would be greatly limited if it were based only on observations made by our unaided senses. Well developed though our senses are, their responses are limited. We cannot hear sound at frequencies above 20,000 Hz. We cannot see electromagnetic radiation outside the limited wavelength region between about 400 nm and 700 nm (1 nm $= 10^{-9}$ m). Even in this visible range, we cannot detect variations in light intensity that occur at a rate faster than about 20 Hz. Although many of the vital processes within our bodies are electrical, our senses cannot detect small electric fields directly. Electrical technology has provided the means for translating information from many areas into the domain of our senses.

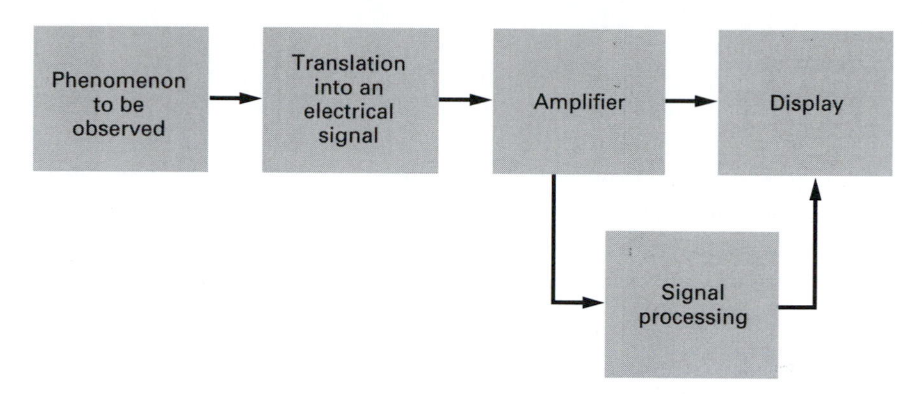

FIGURE 14.1 ▶ An experiment in biology.

Electrical technology is a vast subject that we cannot possibly cover in this short chapter. Here we will simply outline the general techniques used in observing life processes. A description of the various electrical components is found in [14-3] and [14-4].

A diagram of a typical experimental setup in biology is shown in Fig. 14.1. The various subunits of the experiment are shown as blocks of specialized functions. We start with the phenomenon we want to observe but which we cannot detect with our senses. This may be, for example, a high-frequency sound emitted by a bat, the electrochemical activity of a cell, the subtle movement of a muscle, or the light emitted by a fluorescent dye. These phenomena are first translated into electrical signals which then carry information about the intensity and time variations of the original event. Specialized devices are required to perform this task. Some of these devices are commonly found in our everyday technology; others are rather esoteric. Sound, for example, is translated into electrical signals by microphones. Light can be translated into an electrical current by photomultipliers.

The electrical signals generated in this way are usually too weak to drive the final instrument that displays the signals for our observation, so the power and amplitude of the signal are increased by a device called an *amplifier*. The amplified signal then drives the display unit.

The display unit must be matched to the type of signal that is being observed. A slowly varying signal can be displayed on a voltmeter, which has a pointer that moves in accord with the current. Somewhat faster signals are often displayed on a pen recorder, which draws the shape of the signal on a chart. Very fast signals are recorded by a device called an *oscilloscope*. This instrument is similar to a television set. A beam of electrons generated inside the device hits a fluorescent screen that emits lights at the point of impact. The motion and intensity of the beam are controlled by electrical signals applied to the oscilloscope. The resulting picture on the screen displays the information content of the signal. Speakers can also translate into sound electrical signals that have a time variation in the audible range of frequencies.

Often the experimental signal is very noisy. In addition to the desired information, it contains spurious signals due to various sources extraneous to the main phenomenon. Techniques have been developed for analyzing such signals and extracting the relevant information from the noise. In modern experiments, this processing is often performed by a computer.

14.2 Diagnostic Equipment

Most of the diagnostic equipment in medicine utilizes electrical technology in one form or another. Even the traditional stethoscope is now available with electronic modifications that increase its sensitivity. We will describe here only two of the many diagnostic instruments found in a modern clinical facility: the *electrocardiograph* and the *electroencephalograph.*

As a result of the ionic currents associated with electrical activities in the cells, potential differences are produced along the surface of the body. (See Chapter 13.) By measuring these potential differences between appropriate points on the surface of the body, it is possible to obtain information about the functioning of specific organs. The surface potentials are usually very small and, therefore, must be amplified before they can be displayed for examination.

14.2.1 The Electrocardiograph

The electrocardiograph (ECG) is an instrument that records surface potentials associated with the electrical activity of the heart. The surface potentials are conducted to the instrument by metal contacts called *electrodes* which are fixed to various parts of the body. Usually the electrodes are attached to the four limbs and over the heart. Voltages are measured between two electrodes at a time. (See Fig. 14.2.)

A typical normal signal recorded between two electrodes is shown in Fig. 14.3. The main features of this wave form are identified by the letters P, Q, R, S, and T. The shape of these features varies with the location of the electrodes. A trained observer can diagnose abnormalities by recognizing deviations from normal patterns.

The wave shape in Fig. 14.3 is explained in terms of the pumping action of the heart described in Chapter 8. The rhythmic contraction of the heart is initiated by the *pacemaker*, which is a specialized group of muscle cells located near the top of the right atrium. Immediately after the pacemaker fires, the

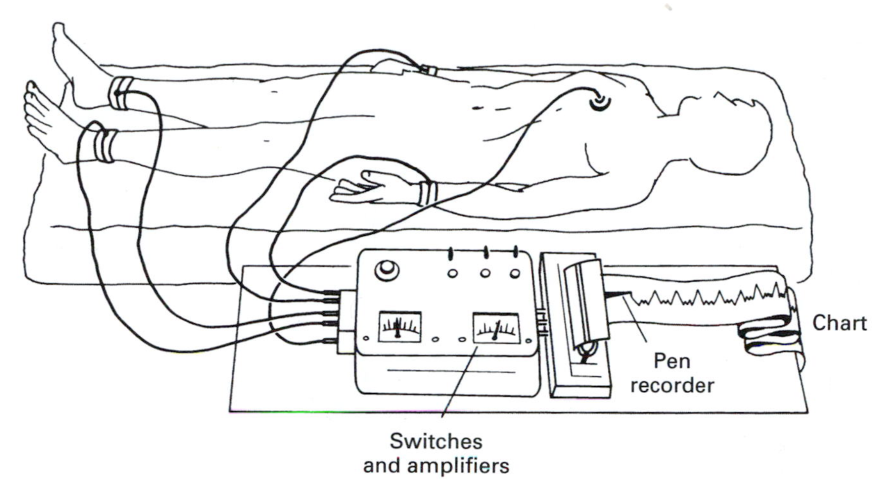

FIGURE 14.2 ▶ Electrocardiography.

action potential propagates through the two atria. The *P* wave is associated with the electrical activity that results in the contraction of the atria. The *QRS* wave is produced by the action potential associated with the contraction of the ventricles. The *T* wave is caused by currents that bring about the recovery of the ventricle for the next cycle.

14.2.2 The Electroencephalograph

The electroencephalograph (EEG) measures potentials along the surface of the scalp. Here again electrodes are attached to the skin at various positions along the scalp. The instrument records the voltages between pairs of electrodes. The EEG signals are much more complex and difficult to interpret

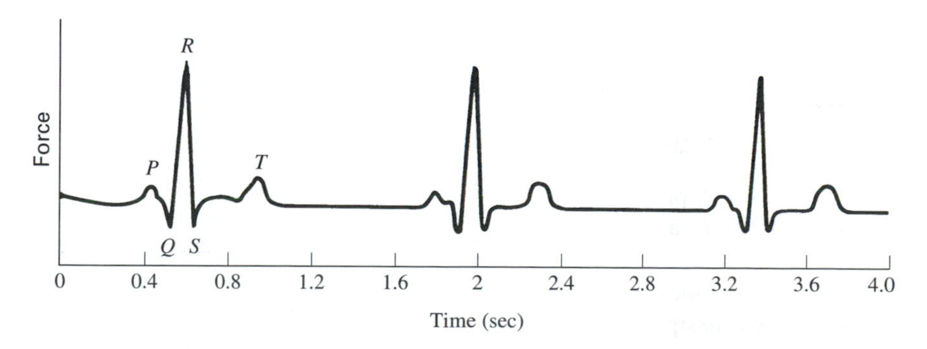

FIGURE 14.3 ▶ An electrocardiogram.

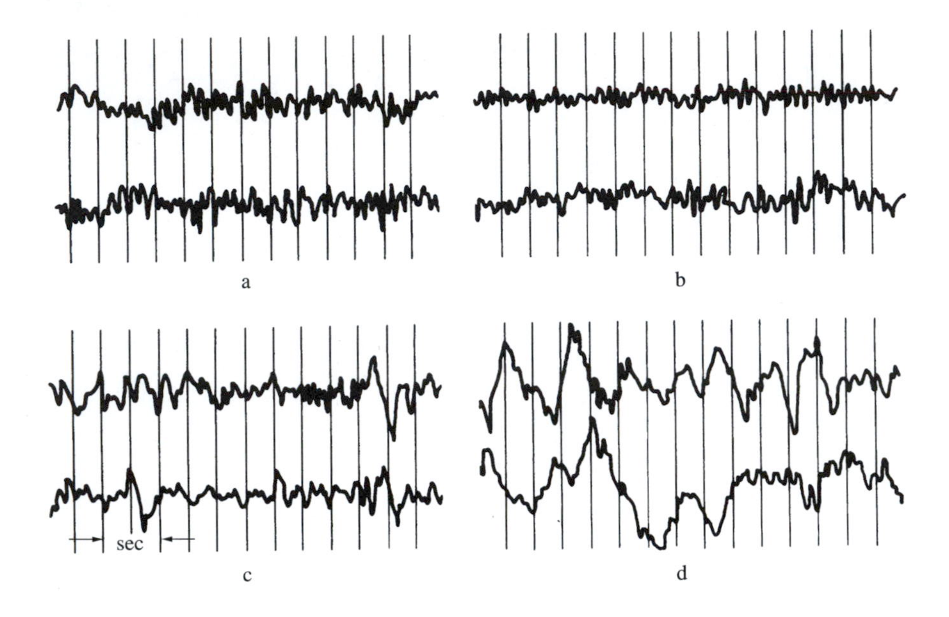

FIGURE 14.4 ▶ EEG potentials between two pairs of electrodes: (a) subject alert, (b) subject drowsy, (c) light sleep, (d) deep sleep.

than those produced by the electrocardiograph. The EEG signals are certainly the result of collective neural activity in the brain. However, so far it has not been possible to relate unambiguously the EEG potentials to specific brain functions. Nevertheless, certain types of patterns are known to be related to specific activities, as illustrated in Fig. 14.4.

Electroencephalographs have been useful in diagnosing various brain disorders. Epileptic seizures, for example, are characterized by pronounced EEG abnormalities (see Fig. 14.5). Brain tumors can often be located by a careful examination of EEG potentials along the whole contour of the scalp.

14.3 Physiological Effects of Electricity

The painful shock produced by electricity is well known to most people. The *shock* results from a current passing through the body. An electrical current has two effects on body tissue. The current stimulates nerves and muscle fibers, which produces pain and a contraction of muscles, and it also heats the tissue through dissipation of electrical energy. Both of these effects, if sufficiently intense, can cause severe injury or death. But if the electrical current

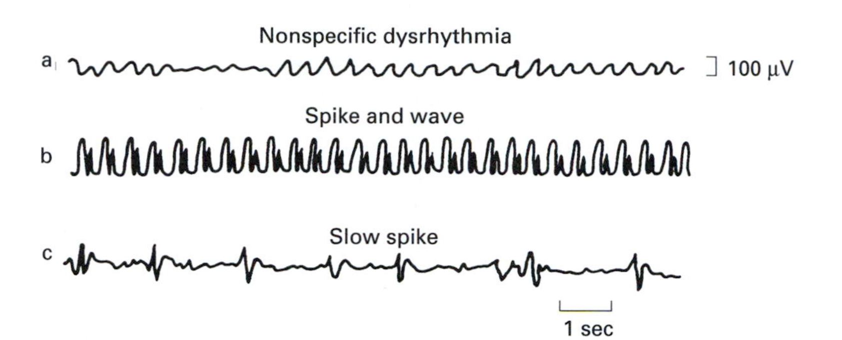

FIGURE 14.5 ▶ Abnormal EEG patterns. Pattern b is typical of petit mal seizures.

is applied in a controlled way, both the heating and the muscle stimulation can be beneficial. For example, local heating of tissue by high-frequency electric currents promotes healing in much the same way as by ultrasonic diathermy.

The amount of current flowing through the body is governed by Ohm's law. Thus, it depends on the voltage of the source and the electrical resistance of the body. The body tissue is a relatively good conductor. Since most of the electrical resistance is in the skin, the danger of electrical shock increases if the skin is wet at the point of contact.

Most people begin to feel an electrical current when it reaches a magnitude of about 500 μA. A 5-mA current causes pain, and currents larger than about 10 mA produce sustained tetanizing contraction of some muscles. This is a dangerous situation because under these conditions the person cannot release the conductor that is delivering the current into his or her body.

The brain, the respiratory muscles, and the heart are all very seriously affected by large electric currents. Currents in the range of a few hundred milliamperes flowing across the head produce convulsions resembling epilepsy. Currents in this range are used in electric shock therapy to treat certain mental disorders.

Currents in the range of a few amperes flowing in the region of the heart can cause death within a few minutes. In this connection, a large current of about 10 A is often less dangerous than a 1-A current. When the smaller current passes through the heart, it may tetanize only part of the heart, thereby causing a desynchronization of the heart action; this condition is called *fibrillation*. The movements of the heart become erratic and ineffective in pumping blood. Usually fibrillation does not stop when the current source is removed. A large current tetanizes the whole heart, and when the current is discontinued the heart may resume its normal rhythmic activity.

Fibrillations often occur during a heart attack and during cardiac surgery. The tetanizing effect of large currents can be used to synchronize the heart. A clinical device designed for this purpose is called a *defibrillator*. A capacitor in this device is charged to about 6000 V and stores about 200 J of energy. Two electrodes connected to the capacitor through a switch are placed on the chest. When the switch is closed, the capacitor rapidly discharges through the body. The current pulse lasts about 5 msec, during which the heart is tetanized (see Exercise 14-1). After the pulse, the heart may resume its normal beat. Often the heart must be shocked a few times before it resynchronizes.

Electric current can also be used to stimulate muscles more gently. We have already mentioned the electric stimulation of paralyzed skeletal muscles to maintain their tone. Heart muscles can be triggered in a similar way. In some heart diseases, the pacemaker cells that control the timing of the heartbeat cease to function properly, and electronic pacemakers have been very useful. The electronic pacemaker is basically a pulse generator that produces short periodic pulses that initiate and control the frequency of the heartbeat. The device can be made small enough for surgical implantation. Unfortunately, the battery that powers the pacemaker has a finite lifetime and must be replaced every few years.

14.4 Control Systems

Many of the processes in living systems must be controlled to meet the requirements of the organism. We have already encountered a few examples of controlled processes in our earlier discussions. Temperature control in the body and the growth of bones were two cases where various processes had to be regulated in order to achieve the desired condition. In this section, we will describe briefly a useful general method of analyzing such control systems.

Features common to all control systems are shown in Fig. 14.6. Each block represents an identifiable function within the control system. The control process consists of:

1. The parameter to be controlled. This may be the temperature of the skin, the movement of muscles, the rate of heart beat, the size of the bone, and so on.

2. A means of monitoring the parameter and transmitting information about its state to some decision-making center. This task is usually performed by the sensory neurons.

3. Some reference value to which the controlled parameters are required to comply. The reference value may be in the central nervous system

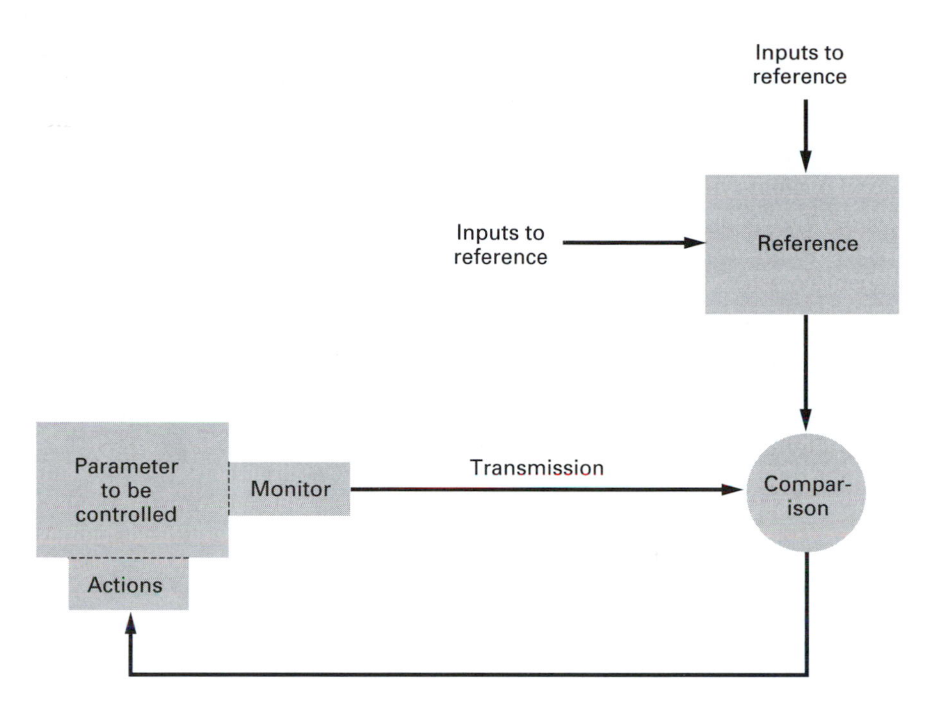

FIGURE 14.6 ► Control of a biological process.

in the form of a decision, for example, about the position of the hand. In this case, the reference value is changeable and is set by the central nervous system. Many references for body functions are autonomous, however, not under the cognitive control of the brain.

4. A method for comparing the state of the parameter with the reference value and for transmitting instructions to bring the two into accord. The instructions may be transmitted by nerve impulses or in some cases by chemical messengers called *hormones*, which diffuse through the body and control various metabolic functions.

5. A mechanism for translating the messages into actions that alter the state of the controlled parameter. In the case of the hand position, for example, this is the contraction of a set of muscle fibers.

We will now illustrate these concepts with a concrete example of the control of the light intensity reaching the retina of the eye (see Fig. 14.7). Light enters the eye through the *pupil*, which is the dark opening in the center of the *iris*. (The iris is the colored disk in the eyeball.) The size of the opening decreases as the light intensity increases. Thus, the iris acts somewhat like the

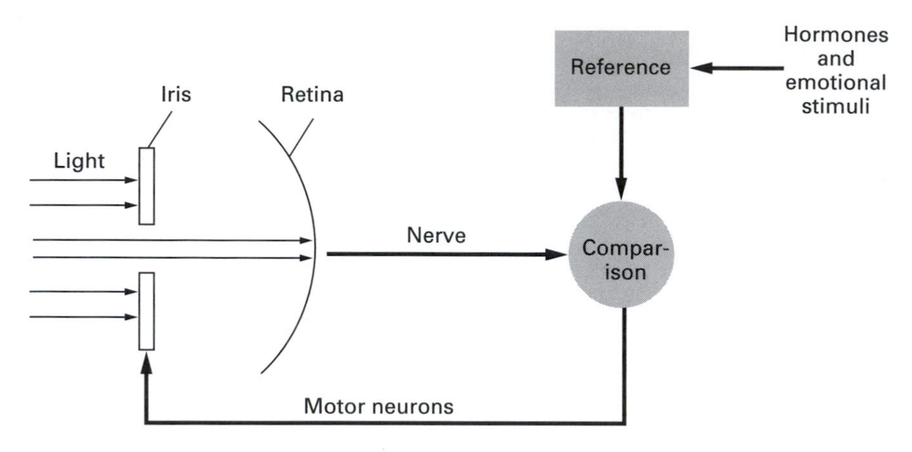

FIGURE 14.7 ▶ Control of the light intensity reaching the retina.

automatic diaphragm in a camera. Clearly this action must be governed by a control system.

Light reaching the retina is converted to neural impulses, which are generated at a frequency proportional to the light intensity. At some place along the nervous system of vision, this information is interpreted and compared to a preset reference value stored probably in the brain. The reference itself can be altered by hormones and various emotional stimuli. The result of this comparison is transmitted by means of nerve impulses to the muscles of the iris which then adjust the size of the opening in response to this signal.

14.5 Feedback

For many years engineers have studied mechanical and electrical systems that have the general characteristics of control systems in biological organisms. Voltage regulators, speed controls, and thermostatic heat regulators all have features in common with biological control systems. Engineers have developed techniques for analyzing and predicting the behavior of control systems. These techniques have also been useful in the study of biological systems.

An engineering analysis of such systems is usually done in terms of input and output. In the light-intensity control example, the input is the light reaching the retina, and the output is the response of the retina to light. The system itself is that which yields an output in response to the input. In our case, this is the retina and the associated nerve circuits. The aim of the iris control system is to maintain the output as constant as possible.

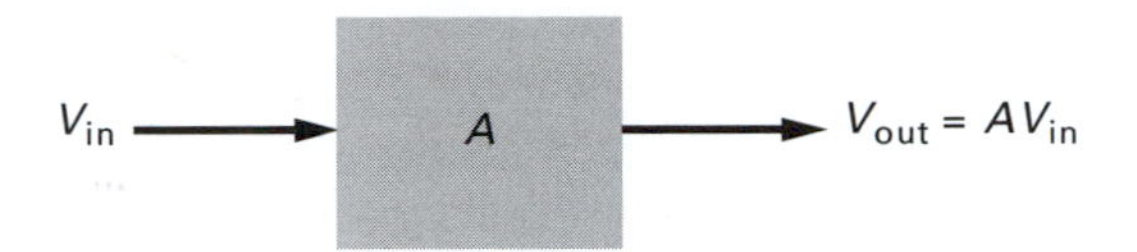

FIGURE 14.8 ▶ An amplifier without feedback.

The most significant point to note about control systems such as the one shown in Fig. 14.7 is that the output affects the input itself. Such systems are called *feedback systems* (because information about the output is fed back to the input). The system is said to have *negative feedback* if it opposes a change in the input and *positive feedback* if it augments a change in the input. The light control in Fig. 14.7 is a negative feedback because an increase in the light intensity causes a decrease in the iris opening and a corresponding reduction of the light reaching the retina. Regulation of body temperature by sweating or shivering is another example of negative feedback, whereas sexual arousal is an example of positive feedback. In general, negative feedback keeps the system response at a relatively constant level. Therefore, most biological feedback systems are in fact negative.

We will illustrate the method of system analysis with an example from electrical engineering. We will analyze in these terms a voltage amplifier that has part of its output fed back to the input. Let us first consider a simple amplifier without feedback (see Fig. 14.8). The amplifier is an electric device that increases the input voltage (V_{in}) by a factor A; that is, the output voltage V_{out} is

$$V_{out} = AV_{in} \tag{14.1}$$

It is evident from this equation that the amplification A is simply determined by the ratio of the output and input voltages; that is,

$$A = \frac{V_{out}}{V_{in}} \tag{14.2}$$

Now let us introduce feedback (Fig. 14.9). Part of the output ($\beta \times V_{out}$) is added back to the amplifier input so that the voltage at the input terminal of

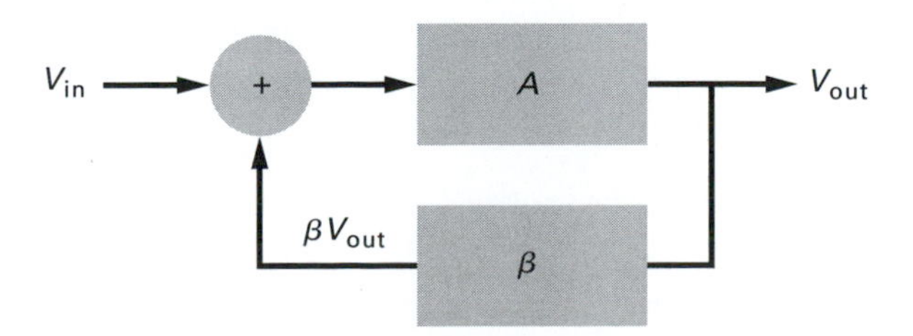

FIGURE 14.9 ▶ An amplifier with feedback.

the amplifier (V_{in}') is

$$V_{in}' = V_{in} + \beta \times V_{out} \tag{14.3}$$

Here V_{in} is the externally applied voltage. The amplification of the total feedback system is

$$A_{feedback} = \frac{V_{out}}{V_{in}} \tag{14.4}$$

Using the fact that $V_{out} = AV_{in}'$, we can show that (see Exercise 14-2)

$$A_{feedback} = \frac{A}{1 - A\beta} \tag{14.5}$$

Now if β is a negative number, the amplification with feedback is smaller than the amplification without feedback (i.e., $A_{feedback}$ is smaller than A). A negative β implies that the voltage is added out of phase with the external input voltage. This is negative feedback. With positive β, we have positive feedback and increased amplification.

This type of analysis has the advantage that we can learn about the system without a detailed knowledge of the individual system components. We can vary the frequency, the magnitude, and the duration of the input voltage and measure the corresponding output voltage. From these measurements, we can obtain some information about the amplifier and the feedback component without knowing anything about the transistors, resistors, capacitors, and other components that make up the device. We could, of course, obtain this information and much more by a detailed analysis of the device in terms of its basic components, but this would involve much more work.

In the study of complex biological functions, the systems approach is often very useful because the details of the various component processes are unknown. For example, in the iris control system, very little is known about the processing of the visual signals, the mechanism of comparing these signals to the reference, or the nature of the reference itself. Yet by shining light at various intensities, wavelengths, and durations into the eye and by measuring the corresponding changes in the iris opening, we can obtain significant information about the system as a whole and even about the various subunits. Here the techniques developed by the engineers are useful in analyzing the system (see Exercises 14-3 and 14-4). However, many biological systems are so complicated with many inputs, outputs, and feedbacks that even the simplified systems approach cannot yield a tractable formulation.

▶ EXERCISES ▶

14-1. From the data in the text, compute the capacitance of the capacitor in the defibrillator and calculate the magnitude of the average current flowing during the pulse.

14-2. Verify Eq. 14.5.

14-3. Draw a block diagram for the following control systems. (a) Control of the body temperature in a person. (b) Control of the hand in drawing a line. (c) Control of the reflex action when the hand draws away from a painful stimulus. Include here the type of control that the brain may exercise on this movement. (d) Control of bone growth in response to pressure.

14-4. For each of the control systems in Exercise 14-3, discuss how the system could be studied experimentally.

Chapter 15

Optics

Light is the electromagnetic radiation in the wavelength region between about 400 and 700 nm (1 nm $= 10^{-9}$ m). Although light is only a tiny part of the electromagnetic spectrum, it has been the subject of much research in both physics and biology. The importance of light is due to its fundamental role in living systems. Most of the electromagnetic radiation from the sun that reaches the Earth's surface is in this region of the spectrum, and life has evolved to utilize it. In photosynthesis, plants use light to convert carbon dioxide and water into organic materials, which are the building blocks of living organisms. Animals have evolved light-sensitive organs which are their main source of information about the surroundings. Some bacteria and insects can even produce light through chemical reactions.

Optics, which is the study of light, is one of the oldest branches of physics. It includes topics such as microscopes, telescopes, vision, color, pigments, illumination, spectroscopy, and lasers, all of which have applications in the life sciences. In this chapter, we will discuss four of these topics: vision, telescopes, microscopes, and optical fibers. Background information needed to understand this chapter is reviewed in Appendix C.

15.1 Vision

Vision is our most important source of information about the external world. It has been estimated that about 70% of a person's sensory input is obtained through the eye. The three components of vision are the stimulus, which is light; the optical components of the eye, which image the light; and the nervous system, which processes and interprets the visual images.

15.2 Nature of Light

Experiments performed during the nineteenth century showed conclusively that light exhibits all the properties of wave motion, which were discussed in Chapter 12. At the beginning of this century, however, it was shown that wave concepts alone do not explain completely the properties of light. In some cases, light and other electromagnetic radiation behave as if composed of small packets (quanta) of energy. These packets of energy are called *photons*. For a given frequency f of the radiation, each photon has a fixed amount of energy E which is

$$E = \hbar f \tag{15.1}$$

where $\hbar$ is Planck's constant, equal to 6.63×10^{-27} erg-sec.

In our discussion of vision, we must be aware of both of these properties of light. The wave properties explain all phenomena associated with the propagation of light through bulk matter, and the quantum nature of light must be invoked to understand the effect of light on the photoreceptors in the retina.

15.3 Structure of the Eye

A diagram of the human eye is given in Fig. 15.1. The eye is roughly a sphere, approximately 2.4 cm in diameter. All vertebrate eyes are similar in structure but vary in size. Light enters the eye through the cornea, which is a transparent section in the outer cover of the eyeball. The light is focused by the lens system of the eye into an inverted image at the photosensitive retina, which covers the back surface of the eye. Here the light produces nerve impulses that convey information to the brain.

The focusing of the light into an image at the retina is produced by the curved surface of the cornea and by the crystalline lens inside the eye. The focusing power of the cornea is fixed. The focus of the crystalline lens, however, is alterable, allowing the eye to view objects over a wide range of distances.

In front of the lens is the iris, which controls the size of the pupil, or entrance aperture into the eye (see Chapter 14). Depending on the intensity of the light, the diameter of the aperture ranges from 2 to 8 mm. The cavity of the eye is filled with two types of fluid, both of which have a refractive index about the same as water. The front of the eye, between the lens and the cornea, is filled with a watery fluid called the *aqueous humor*. The space between the lens and the retina is filled with the gelatinous *vitreous* humor.

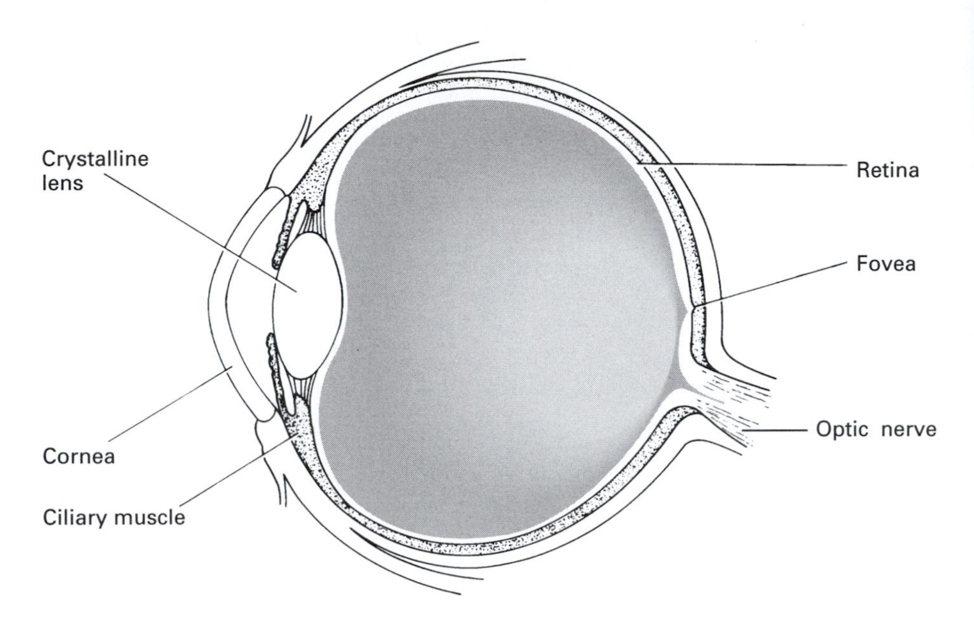

FIGURE 15.1 ▶ The human eye.

15.4 Accommodation

The focusing of the eye is controlled by the ciliary muscle, which can change the thickness and curvature of the lens. This process of focusing is called *accommodation*. When the ciliary muscle is relaxed, the crystalline lens is fairly flat, and the focusing power of the eye is at its minimum. Under these conditions, a parallel beam of light is focused at the retina. Because light from distant objects is nearly parallel, the relaxed eye is focused to view distant objects. In this connection, "distant" is about 6 m and beyond (see Exercise 15-1).

The viewing of closer objects requires greater focusing power. The light from nearby objects is divergent as it enters the eye; therefore, it must be focused more strongly to form an image at the retina. There is, however, a limit to the focusing power of the crystalline lens. With the maximum contraction of the ciliary muscle, a normal eye of a young adult can focus on objects about 15 cm from the eye. Closer objects appear blurred. The minimum distance of sharp focus is called the *near point of the eye*.

The focusing range of the crystalline lens decreases with age. The near point for a 10-year-old child is about 7 cm, but by the age of 40 the near point shifts to about 22 cm. After that the deterioration is rapid. At age 60, the near point is shifted to about 100 cm. This decrease in the accommodation of the eye with age is called *presbyopia*.

15.5 Eye and the Camera

Although the designers of the photographic camera did not consciously imitate the structure of the eye, many of the features in the two are remarkably similar (see Fig. 15.2). Both consist of a lens system that focuses a real inverted image onto a photosensitive surface. In the eye, as in the camera, the diameter of the light entrance is controlled by a diaphragm that is adjusted in accord with the available light intensity. In a camera, the image is focused by moving the lens with respect to the film. In the eye, the distance between the retina and the lens is fixed; the image is focused by changing the thickness of the lens.

Even the photosensitive surfaces are somewhat similar. Both photographic film and the retina consist of discrete light-sensitive units, microscopic in size, which undergo chemical changes when they are illuminated. In fact, under special circumstances, the retina can be "developed," like film, to show the image that was projected on it. This was first demonstrated in the 1870s by the German physiologist W. Kuhne. He exposed the eye of a living rabbit to light coming through a barred window. After 3 minutes of exposure to light, the rabbit was killed and its retina was immersed in an alum solution which fixed the retinal reaction. The barred window was clearly visible on the retina. A few years later, Kuhne fixed the retina from the head of a guillotined criminal. He observed an image, but he could not interpret it in terms of anything that the man had seen before he was beheaded.

The analogy between the eye and the camera, however, is not complete.

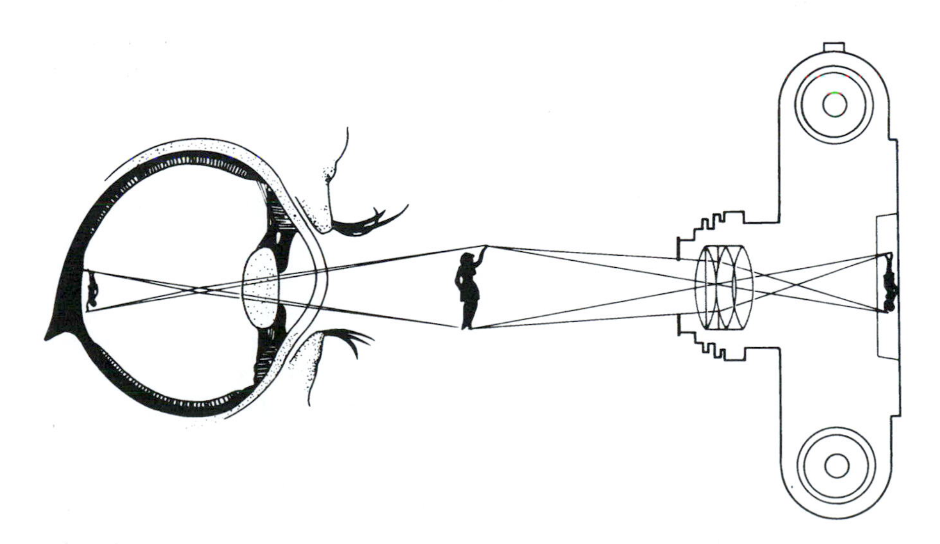

FIGURE 15.2 ▶ The eye and the camera.

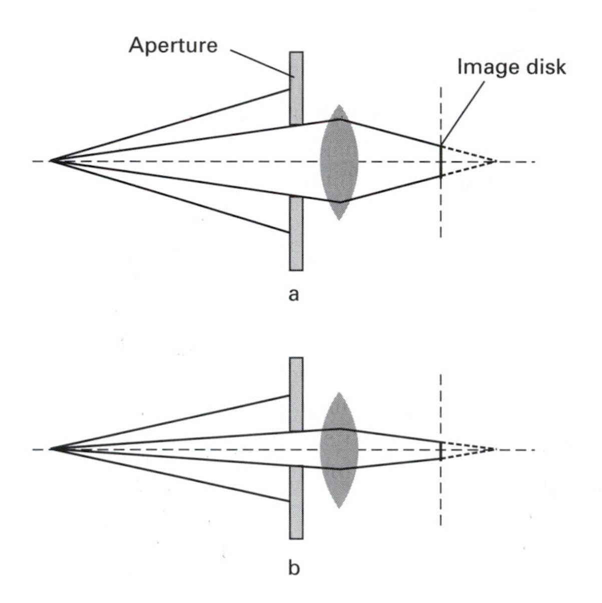

FIGURE 15.3 ▶ Size of image disk: (a) with large aperture, (b) with small aperture.

As we will describe later, the eye goes far beyond the camera in processing the images that are projected on the retina.

15.5.1 Aperture and Depth of Field

The iris is the optical aperture of the eye, and its size varies in accordance with the available light. If there is adequate light, the quality of the image is best with the smallest possible aperture. This is true for both the eye and the camera.

There are two main reasons for the improved image with reduced aperture. Imperfections in lenses tend to be most pronounced around the edges. A small aperture restricts the light path to the center of the lens and eliminates the distortions and aberrations produced by the periphery.

A smaller aperture also improves the image quality of objects that are not located at the point on which the eye or the camera is focused. An image is in sharp focus at the retina (or film) only for objects at a specific distance from the lens system. Images of objects not at this specific plane are blurred at the retina (see Fig. 15.3); in other words, a point that is not in exact focus appears as a disk on the retina. The amount of blurring depends on the size of the aperture. As shown in Fig. 15.3, a small aperture reduces the diameter of the blurred spot and allows the formation of a relatively clear image from objects that are not on the plane to which the eye is focused. The range of

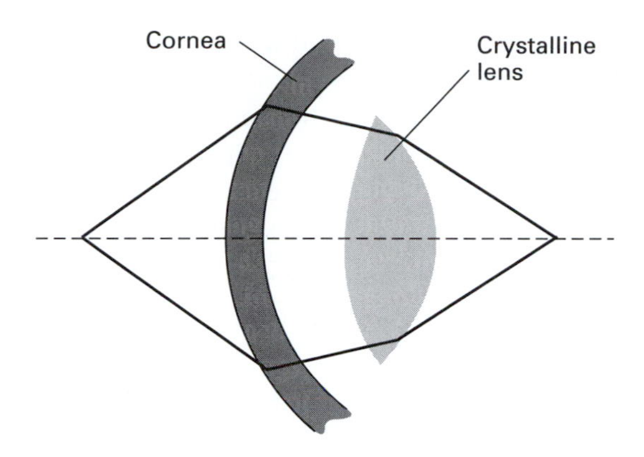

FIGURE 15.4 ▶ Focusing by the cornea and the crystalline lens (not to scale).

object distances over which a good image is formed for a given setting of the focus is called the *depth of field*. Clearly a small aperture increases the depth of field. It can be shown that the depth of field is inversely proportional to the diameter of the aperture (see Exercise 15-2).

15.6 Lens System of the Eye

The focusing of the light into a real inverted image at the retina is produced by refraction at the cornea and at the crystalline lens (see Fig. 15.4). The focusing or refractive power of the cornea and the lens can be calculated using Eq. C.9, (Appendix C). The data required for the calculation are shown in Table 15.1.

The largest part of the focusing, about two thirds, occurs at the cornea. The power of the crystalline lens is small because its index of refraction is only slightly greater than that of the surrounding fluid. In Exercise 15-3, it is shown that the refractive power of the cornea is 42 diopters, and the refractive power of the crystalline lens is variable between about 19 and 24 diopters. (For a definition of the unit *diopter*, see Appendix C.)

TABLE 15.1 ▶ **Parameters for the Eye**

	Radius (mm)		Index of
	Front	Back	refraction
Cornea	7.8	7.3	1.38
Lens, min. power	10.00	−6.0	
Lens, max. power	6.0	−5.5	1.40
Aqueous and vitreous humor			1.33

The refractive power of the cornea is greatly reduced when it is in contact with water (see Exercise 15-4). Because the crystalline lens in the human eye cannot compensate for the diminished power of the cornea, the human eye under water is not able to form a clear image at the retina and vision is blurred. In fish eyes, which have evolved for seeing under water, the lens is intended to do most of the focusing. The lens is nearly spherical and has a much greater focusing power than the lens in the eyes of terrestrial animals (see Exercise 15-5).

15.7 Reduced Eye

To trace accurately the path of a light ray through the eye, we must calculate the refraction at four surfaces (two at the cornea and two at the lens). It is possible to simplify this laborious procedure with a model called the *reduced eye*, shown in Fig. 15.5. Here all the refraction is assumed to occur at the front surface of the cornea, which is constructed to have a diameter of 5 mm. The eye is assumed to be homogeneous, with an index of refraction of 1.333 (the same as water). The retina is located 2 cm behind the cornea. The nodal point n is the center of corneal curvature located 5 mm behind the cornea.

This model represents most closely the relaxed eye which focuses parallel light at the retina, as can be confirmed using Eq. C.9. For the reduced eye, the second term on the right-hand side of the equation vanishes because the light is focused within the reduced eye so that $n_L = n_2$. Equation C.9, therefore, simplifies to

$$\frac{n_1}{p} + \frac{n_L}{q} = \frac{n_L - n_1}{R} \tag{15.2}$$

where $n_1 = 1, n_L = 1.333$, and $R = 0.5$ cm. Because the incoming light is parallel, its source is considered to be at infinity (i.e., $p = \infty$). Therefore, the

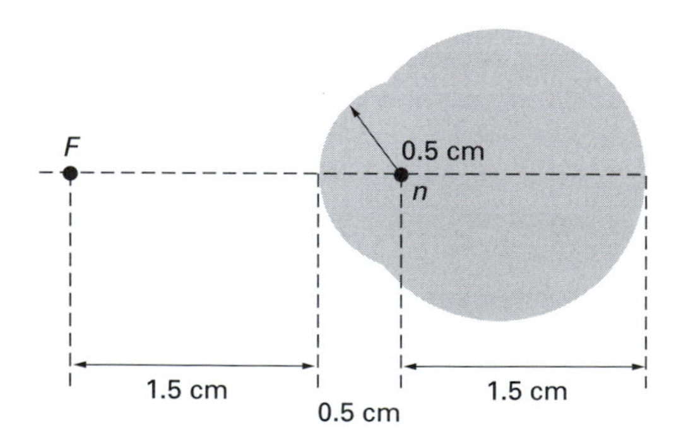

FIGURE 15.5 ▶ The reduced eye.

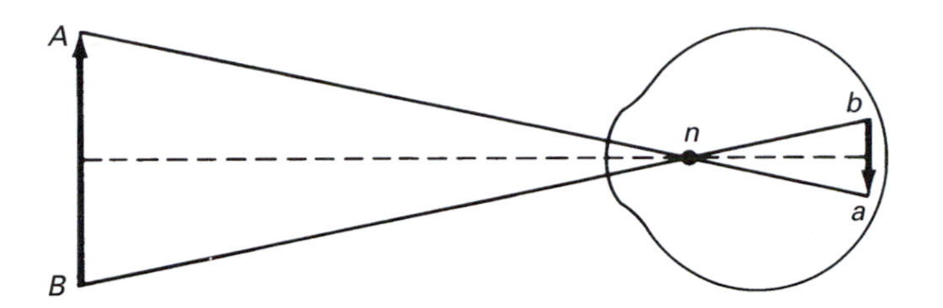

FIGURE 15.6 ▶ Determination of the image size on the retina.

distance q at which parallel light is focused is given by

$$\frac{1.333}{q} = \frac{1.333 - 1}{5}$$

or

$$q = \frac{1.333 \times 5}{0.333} = 20 \text{ mm}$$

The anterior focal point F for the reduced eye is located 15 mm in front of the cornea. This is the point at which parallel light originating inside the eye is focused when it emerges from the eye (see Exercise 15-6).

Although the reduced eye does not contain explicitly the mechanism of accommodation, we can use the model to determine the size of the image formed on the retina. The construction of such an image is shown in Fig. 15.6. Rays from the limiting points of the object A and B are projected through the nodal point to the retina. The limiting points of the image at the retina are a and b. This construction assumes that all the rays from points A and B that enter the eye are focused on the retina at points a and b, respectively. Rays from all other points on the object are focused correspondingly between these limits. The triangles AnB and anb are similar; therefore, the relation of object to image size is given by

$$\frac{\text{Object size}}{\text{Image size}} = \frac{\text{Distance of object from nodal point}}{\text{Distance of image from nodal point}} \tag{15.3}$$

or

$$\frac{AB}{ab} = \frac{An}{an}$$

Consider as an example the image of a person 180 cm tall standing 2 m from the eye. The height of the full image at the retina is

$$\text{Height of image} = 180 \times \frac{1.5}{205} = 1.32 \text{ cm}$$

The size of the face in the image is about 1.8 mm, and the nose is about 0.4 mm.

15.8 Retina

The retina consists of photoreceptor cells in contact with a complex network of neurons and nerve fibers which are connected to the brain via the optic nerve (see Fig. 15.7). Light absorbed by the photoreceptors produces nerve impulses that travel along the neural network and then through the optic nerve into the brain. The photoreceptors are located behind the neural network, so the light must pass through this cell layer before it reaches the photoreceptors.

There are two types of photoreceptor cells in the retina: *cones* and *rods*. The cones are responsible for sharp color vision in daylight. The rods provide vision in dim light.

Near the center of the retina is a small depression about 0.3 mm in diameter which is called the *fovea*. It consists entirely of cones packed closely together. Each cone is about 0.002 mm (2 μm) in diameter. Most detailed vision is obtained on the part of the image that is projected on the fovea. When the eye scans a scene, it projects the region of greatest interest onto the fovea.

The region around the fovea contains both cones and rods. The structure of the retina becomes more coarse away from the fovea. The proportion of cones decreases until, near the edge, the retina is composed entirely of rods. In the fovea, each cone has its own path to the optic nerve. This allows the

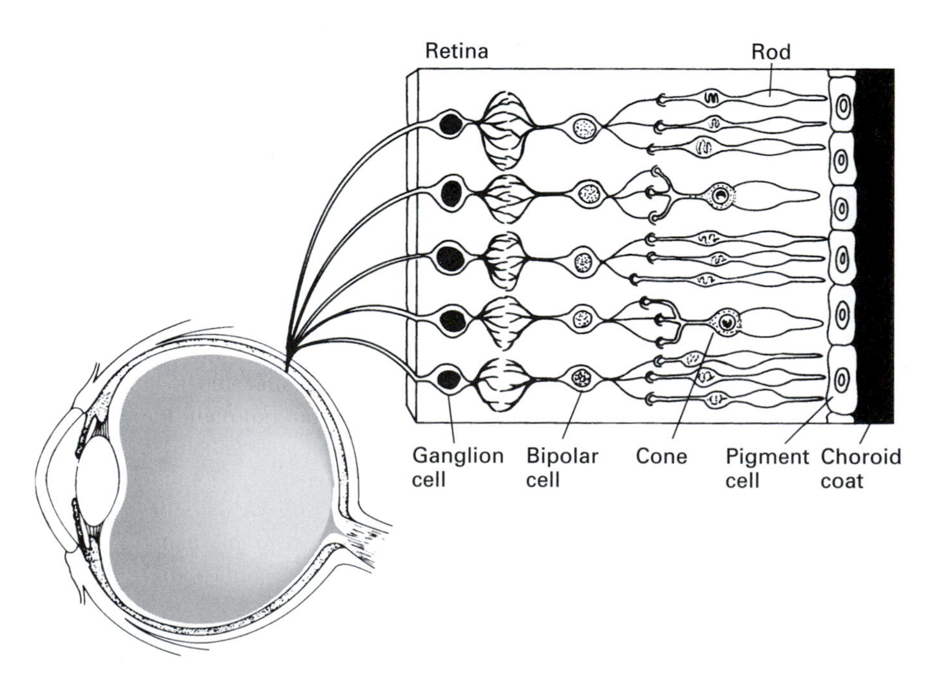

FIGURE 15.7 ▶ The retina.

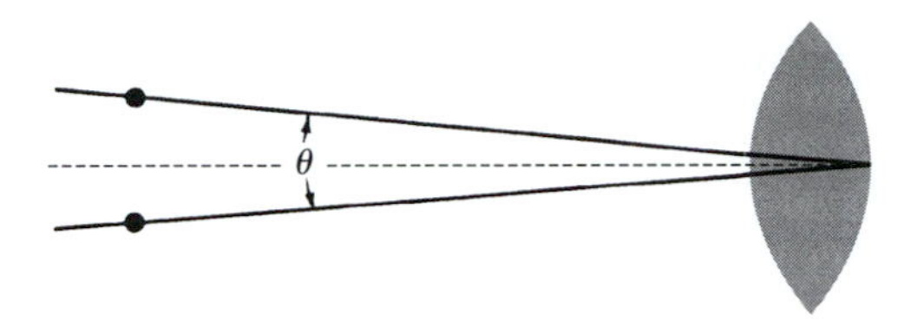

FIGURE 15.8 ▶ Two points are resolvable if the angle θ is greater than $1.22\lambda/d$.

perception of details in the image projected on the fovea. Away from the fovea, a number of receptors are attached to the same nerve path. Here the resolution decreases, but the sensitivity to light and movement increases.

With the structure of the retina in mind, let us examine how we view a scene from a distance of about 2 m. From this distance, at any one instant, we can see most distinctly an object only about 4 cm in diameter. An object of this size is projected into an image about the size of the fovea.

Objects about 20 cm in diameter are seen clearly but not with complete sharpness. The periphery of large objects appears progressively less distinct. Thus, for example, if we focus on a person's face 2 m away, we can see clearly the facial details, but we can pick out most clearly only a subsection about the size of the mouth. At the same time, we are aware of the person's arms and legs, but we cannot detect, for example, details about the person's shoes.

15.9 Resolving Power of the Eye

So far in our discussion of image formation we have used geometric optics, which neglects the diffraction of light. Geometric optics assumes that light from a point source is focused into a point image. This is not the case. When light passes through an aperture such as the iris, diffraction occurs, and the wave spreads around the edges of the aperture.[1] As a result, light is not focused into a sharp point but into a diffraction pattern consisting of a disk surrounded by rings of diminishing intensity.

If light originates from two point sources that are close together, their image diffraction disks may overlap, making it impossible to distinguish the two points. An optical system can resolve two points if their corresponding diffraction patterns are distinguishable. This criterion alone predicts that two points are resolvable (see Fig. 15.8) if the angular separation between the lines

[1] If there are no smaller apertures in the optical path, the lens itself must be considered as the aperture.

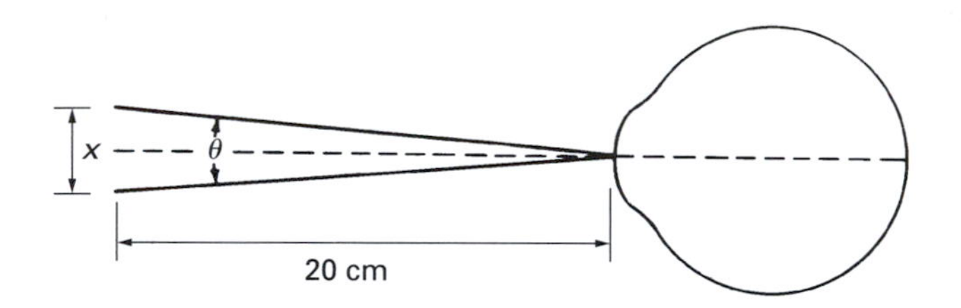

FIGURE 15.9 ▶ Resolution of the eye.

joining the points to the center of the lens is equal to or greater than a critical value given by

$$\theta = \frac{1.22\lambda}{d} \tag{15.4}$$

where λ is the wavelength of light and d is the diameter of the aperture. The angle θ is given in radians (1 rad = 57.3°). With green light (λ = 500 nm) and an iris diameter of 0.5 cm, this angle is 1.22×10^{-4} rad.

Experiments have shown that the eye does not perform this well. Most people cannot resolve two points with an angular separation of less than 5×10^{-4} rad. Clearly there are other factors that limit the resolution of the eye. Imperfections in the lens system of the eye certainly impede the resolution. But perhaps even more important are the limitations imposed by the structure of the retina.

The cones in the closely packed fovea are about 2 μm diameter. To resolve two points, the light from each point must be focused on a different cone and the excited cones must be separated from each other by at least one cone that is not excited. Thus at the retina, the images of two resolved points are separated by at least 4μm. A single unexcited cone between points of excitation implies an angular resolution of about 3×10^{-4} rad (see Exercise 15-7a). Some people with acute vision do resolve points with this separation, but most people do not. We can explain the limits of resolution demonstrated by most normal eyes if we assume that, to perceive distinct point images, there must be three unexcited cones between the areas of excitation. The angular resolution is then, as observed, 5×10^{-4} rad (see Exercise 15-7b).

Let us now calculate the size of the smallest detail that the unaided eye can resolve. To observe the smallest detail, the object must be brought to the closest point on which the eye can focus. Assuming that this distance is 20 cm from the eye, the angle subtended by two points separated by a distance x is given by (see Fig. 15.9)

$$\tan^{-1}\frac{\theta}{2} = \frac{x/2}{20} \tag{15.5}$$

If θ is very small, as is the case in our problem, the tangent of the angle is equal to the angle itself and

$$\theta = \frac{x}{20}$$

Because the smallest resolvable angle is 5×10^{-4} rad, the smallest resolvable detail x is

$$x = 5 \times 10^{-4} \times 20 = 100\ \mu\text{m} = 0.1\ \text{mm}$$

Using the same criterion, we can show (see Exercise 15-8) that the facial features such as the whites of the eye are resolvable from as far as 20 m.

15.10 Threshold of Vision

The sensation of vision occurs when light is absorbed by the photosensitive rods and cones. At low levels of light, the main photoreceptors are the rods. Light produces chemical changes in the photoreceptors which reduce their sensitivity. For maximum sensitivity the eye must be kept in the dark (dark adapted) for about 30 minutes to restore the composition of the photoreceptors.

Under optimum conditions, the eye is a very sensitive detector of light. The human eye, for example, responds to light from a candle as far away as 20 km. At the threshold of vision, the light intensity is so small that we must describe it in terms of photons. Experiments indicate that an individual photoreceptor (rod) is sensitive to 1 quantum of light. This, however, does not mean that the eye can see a single photon incident on the cornea. At such low levels of light, the process of vision is statistical.

In fact, measurements show that about 60 quanta must arrive at the cornea for the eye to perceive a flash. Approximately half the light is absorbed or reflected by the ocular medium. The 30 or so photons reaching the retina are spread over an area containing about 500 rods. It is estimated that only 5 of these photons are actually absorbed by the rods. It seems, therefore, that at least 5 photoreceptors must be stimulated to perceive light.

The energy in a single photon is very small. For green light at 500 nm, it is

$$E = hf = \frac{hc}{\lambda} = \frac{6.63 \times 10^{-27} \times 3 \times 10^{10}}{5 \times 10^{-5}} = 3.98 \times 10^{-12}\ \text{erg}$$

This amount of energy, however, is sufficient to initiate a chemical change in a single molecule which then triggers the sequence of events that leads to the generation of the nervous impulse.

15.11 Vision and the Nervous System

Vision cannot be explained entirely by the physical optics of the eye. There are many more photoreceptors in the retina than fibers in the optic nerve. It is, therefore, evident that the image projected on the retina is not simply transmitted point by point to the brain. A considerable amount of signal processing occurs in the neural network of the retina before the signals are transmitted to the brain. The neural network "decides" which aspects of the image are most important and stresses the transmission of those features. In a frog, for example, the neurons in the retina are organized for most active response to movements of small objects. A fly moving across the frog's field of vision will produce an intense neural response, and if the fly is close enough, the frog will lash out its tongue to capture the fly. On the other hand, a large object, clearly not food for the frog, moving in the same vision field will not elicit a neural response. Evidently the optical processing system of the frog enhances its ability to catch small insects while reducing the likelihood of being noticed by larger, possibly dangerous creatures passing through the neighborhood.

The human eye also possesses important processing mechanisms. It has been shown that movement of the image is necessary for human vision as well. In the process of viewing an object, the eye executes small rapid movements, 30 to 70 per second, which alter slightly the position of the image on the retina. Under experimental conditions, it is possible to counteract the movement of the eye and stabilize the position of the retinal image. It has been found that, under these conditions, the image perceived by the person gradually fades.

15.12 Defects in Vision

There are three common defects in vision associated with the focusing system of the eye: *myopia* (nearsightedness), *hyperopia* (farsightedness), and *astigmatism*. The first two of these defects are best explained by examining the imaging of parallel light by the eye.

The relaxed normal eye focuses parallel light onto the retina (Fig. 15.10). In the myopic eye the lens system focuses the parallel light in front of the retina (Fig. 15.11a). This misfocusing is usually caused by an elongated eyeball or an excessive curvature of the cornea. In hyperopia the problem is reversed (see Fig. 15.12a). Parallel light is focused behind the retina. The problem here is caused by an eyeball that is shorter than normal or by the inadequate focusing power of the eye. The hyperopic eye can accommodate objects at infinity, but its near point is farther away than is normal. Hyperopia is, thus, similar to presbyopia. These two defects can be summarized as fol-

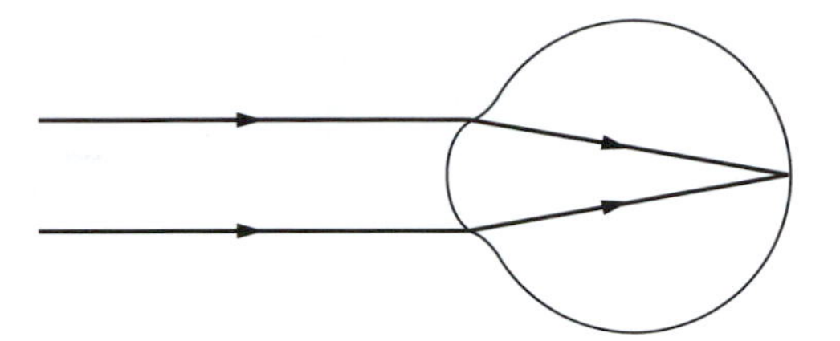

FIGURE 15.10 ▶ The normal eye.

lows: The myopic eye converges light too much, and the hyperopic eye not enough.

Astigmatism is a defect caused by a nonspherical cornea. An oval-shaped cornea, for example, is more sharply curved along one plane than another; therefore, it cannot form simultaneously sharp images of two perpendicular lines. One of the lines is always out of focus, resulting in distorted vision.

All three of these defects can be corrected by lenses placed in front of the eye. Myopia requires a diverging lens to compensate for the excess refraction in the eye. Hyperopia is corrected by a converging lens, which adds to the focusing power of the eye. The uneven corneal curvature in astigmatism is compensated for by a cylindrical lens (Fig. 15.13), which focuses light along one axis but not along the other.

15.13 Lens for Myopia

Let us assume that the farthest object a certain myopic eye can properly focus is 2 m from the eye. This is called the *far point of the eye*. Light from objects farther away than this is focused in front of the retina (Fig. 15.11a). Here the purpose of the corrective lens is to make parallel light appear to come from the far point of the eye (in this case, 2 m). With such a corrective lens, the eye is able to form images of objects all the way to infinity.

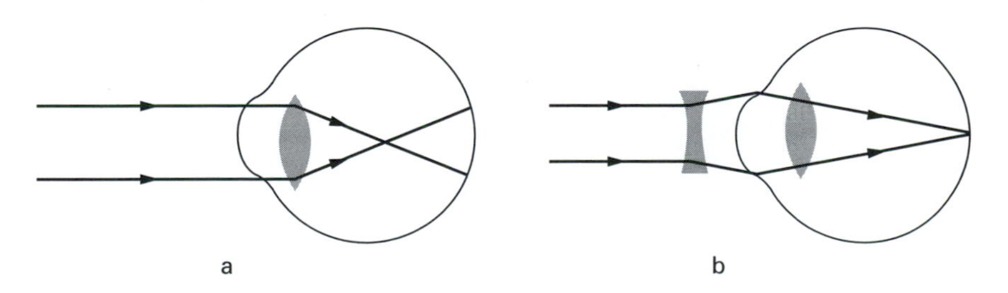

FIGURE 15.11 ▶ (a) Myopia. (b) Its correction.

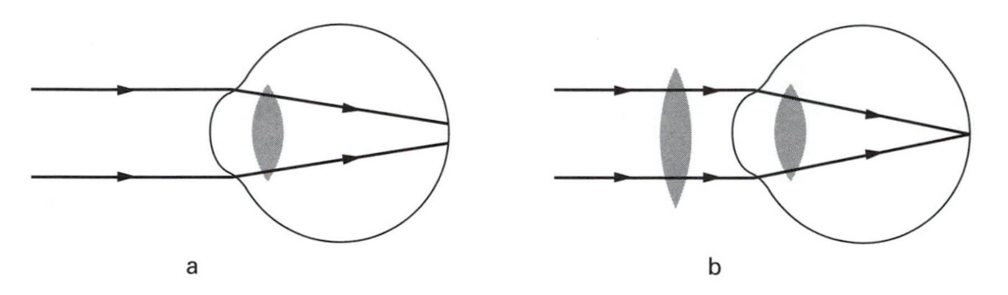

FIGURE 15.12 ▶ (a) Hyperopia. (b) Its correction.

The focal length of the lens is obtained by using Eq. C.6, which is

$$\frac{1}{p} + \frac{1}{q} = \frac{1}{f}$$

Here p is infinity, as this is the effective distance for sources of parallel light. The desired location q for the virtual image is -200 cm. The focal length of the diverging lens (see Eq. C.4) is, therefore,

$$\frac{1}{f} = \frac{1}{\infty} + \frac{1}{-200} \qquad \text{or} \qquad f = -200 \text{ cm} = -5 \text{ diopters}$$

15.14 Lens for Presbyopia and Hyperopia

In these disorders, the eye cannot focus properly on close objects. The near point is too far from the eye. The purpose of the lens is to make light from close objects appear to come from the near point of the unaided eye. Let us assume that a given hyperopic eye has a near point at 150 cm. The desired lens is to allow the eye to view objects at 25 cm. The focal length of the lens is again obtained from Eq. C.6,

$$\frac{1}{p} + \frac{1}{q} = \frac{1}{f}$$

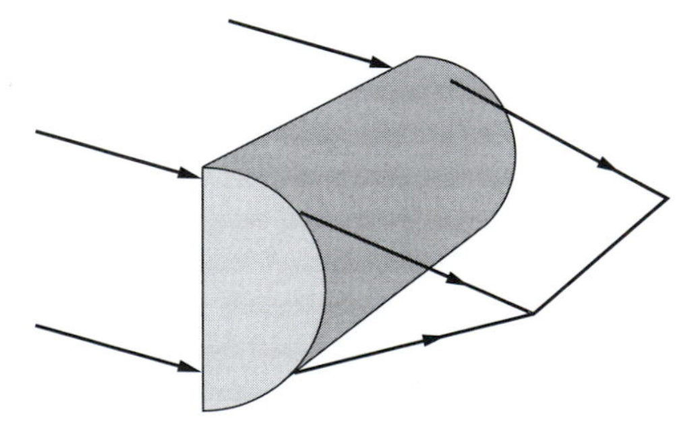

FIGURE 15.13 ▶ Cylindrical lens for astigmatism.

Here p is the object distance at 25 cm and q is -150 cm, which is the distance of the virtual image at the near point. The focal length f for the converging lens is given by

$$\frac{1}{f} = \frac{1}{25\text{ cm}} - \frac{1}{150\text{ cm}} \qquad \text{or} \qquad f = 30\text{ cm} = 33.3\text{ diopters}$$

15.15 Extension of Vision

The range of vision of the eye is limited. Details on distant objects cannot be seen because their images on the retina are too small. The retinal image of a 20 m-high tree at a distance of 500 m is only 0.6 mm high. The leaves on this tree cannot be resolved by the unaided eye (see Exercise 15-9). Observation of small objects is limited by the accommodation power of the eye. We have already shown that, because the average eye cannot focus light from objects closer than about 20 cm, its resolution is limited to approximately 100μm.

Over the past 300 years, two types of optical instruments have been developed to extend the range of vision: the telescope and the microscope. The telescope is designed for the observation of distant objects. The microscope is used to observe small objects that cannot be seen clearly by the naked eye. Both of these instruments are based on the magnifying properties of lenses. A third more recent aid to vision is the fiberscope which utilizes total internal reflection to allow the visualization of objects normally hidden from view.

15.15.1 Telescope

A drawing of a simple telescope is shown in Fig. 15.14. Parallel light from a distant object enters the first lens, called the *objective lens* or *objective*, which forms a real inverted image of the distant object. Because light from

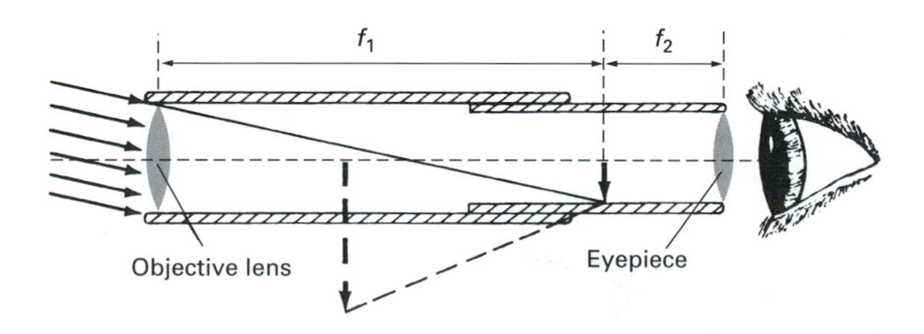

FIGURE 15.14 ▶ The telescope.

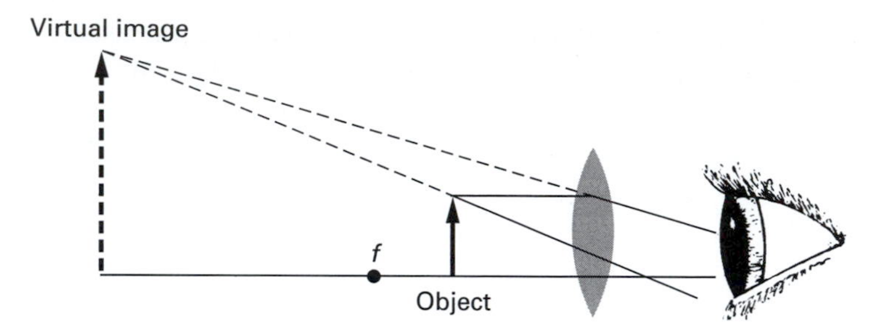

FIGURE 15.15 ▶ Simple magnifier.

the distant object is nearly parallel, the image is formed at the focal plane of the objective. (The drawing shows the light rays from only a single point on the object.) The second lens, called the *eyepiece*, magnifies the real image. The telescope is adjusted so that the real image formed by the objective falls just within the focal plane of the eyepiece. The eye views the magnified virtual image formed by the eyepiece. The total magnification—the ratio of image to object size—is given by

$$\text{Magnification} = -\frac{f_1}{f_2} \tag{15.6}$$

where f_1 and f_2 are the focal lengths of the objective and the eyepiece respectively. As can be seen from Eq. 15.6, greatest magnification is obtained with a long focal-length objective and a short focal-length eyepiece.

15.15.2 Microscope

A simple microscope consists of a single lens that magnifies the object (Fig. 15.15). Better results can be obtained, however, with a two-lens system compound microscope, shown in Fig. 15.16. The compound microscope,

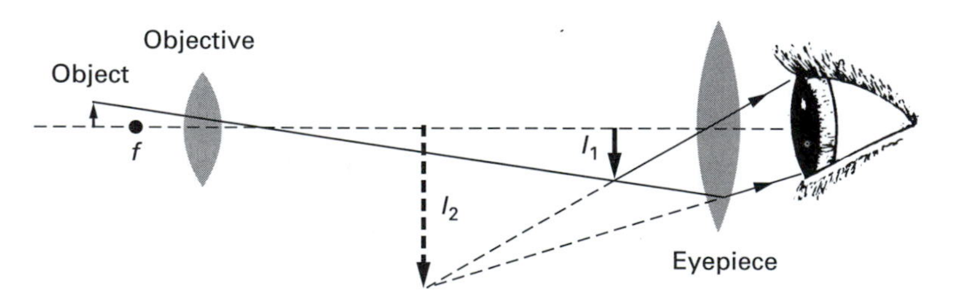

FIGURE 15.16 ▶ Schematic diagram of a compound microscope.

like the telescope, consists of an objective lens and an eyepiece, but the objective of the microscope has a short focal length. It forms a real image I_1 of the object; the eye views the final magnified image I_2 formed by the eyepiece.

The microscope is an important tool in the life sciences. Its invention in the 1600s marked the beginning of the study of life on the cellular level. The early microscope produced highly distorted images, but years of development perfected the device nearly to its theoretical optimum. The resolution of the best modern microscopes is determined by the diffraction properties of light, which limit the resolution to about half the wavelength of light. In other words, with a good modern microscope, we can observe objects as small as half the wavelength of the illuminating light.

We will not present here the details of microscopy. These can be found in many basic physics texts (see, for example, [15.1]). We will, however, describe a special-purpose scanning confocal microscope designed in our laboratory by Paul Davidovits and M. David Egger.

15.15.3 Confocal Microscopy

With conventional microscopes, it is not possible to observe small objects embedded in translucent materials. For example, cells located beneath the surface of tissue, such as buried brain cells in living animals, cannot be satisfactorily observed with conventional microscopes.

Light can certainly penetrate through tissue. This can be demonstrated simply by inserting a flashlight into the mouth and observing the light passing through the cheeks. In principle, therefore, we should be able to form a magnified image of a cell inside the tissue. This could be done by shining light into the tissue and collecting the light reflected from the cell. Unfortunately there is a problem associated with the straightforward use of this technique. Light is reflected and scattered not only by the cell of interest but also by the surface of the tissue and by the cells in front and behind the cell being examined. This spurious light is also intercepted by the microscope and masks the image of the single cell layer within the tissue (see Fig. 15.17).

Over the years, a number of microscopes have been designed that attempted to solve this problem. The most successful of these is the *confocal microscope*. The principle of confocal microscopy was first described by Marvin Minsky in 1957. In the 1960s, Davidovits and Egger modified the Minsky design and built the first successful confocal microscope for observation of cells within living tissue.

The confocal microscope is designed to accept light only from a thin slice within the tissue and to reject light reflected and scattered from other regions. A schematic diagram of the Davidovits-Egger microscope is shown in Fig. 15.18. Although the device does not resemble a conventional micro-

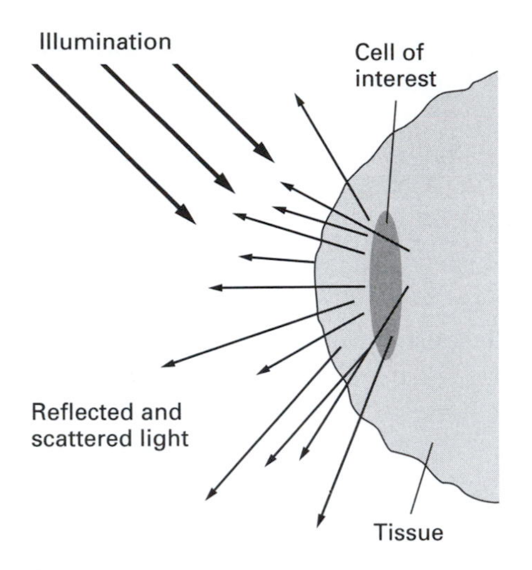

FIGURE 15.17 ▶ Light scattered and reflected from tissue.

scope, it certainly does produce magnified images. This microscope requires a parallel beam of light for illumination of the object. As the source of parallel light we used a laser with a power output that is relatively low so that it does not damage the tissue under observation. The laser beam is reflected by a half-silvered mirror into the objective lens, which focuses the beam to a point inside the tissue. Because the light is parallel, the beam is brought to a point at the principal focus of the lens. The depth of this point in the tissue can be changed by altering the distance between the lens and the tissue.

Light is scattered and reflected from all points in the path of the entering light, and part of this returning light is intercepted by the objective lens. However, only light originating from the focal point emerges from the lens as a parallel beam; light from all other points either converges toward or diverges from the lens axis. The returning light passes through the half-silvered mirror and is intercepted by the collecting lens. Only the parallel component of the light is focused into the small exit aperture that is placed at the principal focal point of the collecting lens. Nonparallel light is defocused at the exit aperture. A photomultiplier placed behind the exit aperture produces a voltage proportional to the light intensity transmitted through the exit aperture. This voltage is then used to control the intensity of an electron beam in the oscilloscope.

So far, we have one spot on the screen of the oscilloscope which glows with a brightness proportional to the reflectivity of one point inside the tissue. In order to see a whole cell or region of cells, we must scan the region point by point. This is done by moving the lens in its own plane so that the focal point scans an area inside the tissue. The motion of the lens does not

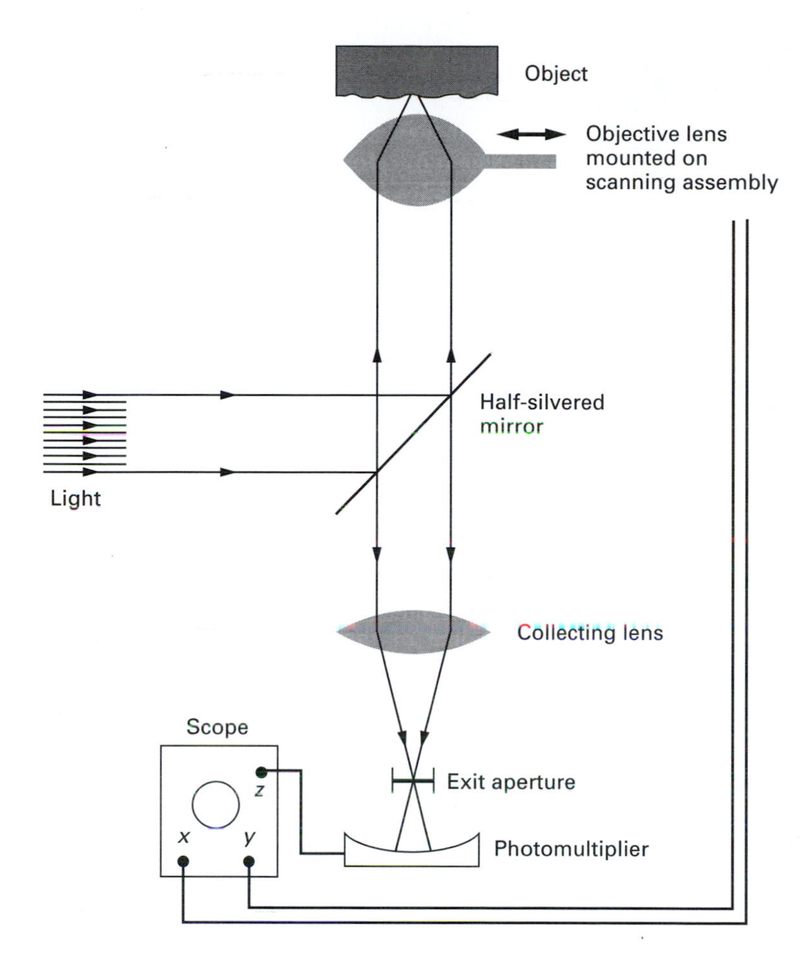

FIGURE 15.18 ▶ Confocal microscope.

affect the parallelism of the light originating at the focal point of the objective lens. Therefore, at every instant, the output of the photomultiplier and the corresponding brightness of the spot on the screen are proportional to the reflectivity of the point being scanned. While the object is scanned, the electron beam in the oscilloscope is moved in synchrony with the motion of the objective lens. Thus, the screen shows a picture of a very thin section within the tissue.

The magnification of this microscope is simply the ratio of the electron beam excursion on the oscilloscope face to the excursion of the scanning lens. For a 0.1-mm excursion of the lens, the electron beam may be adjusted to move 5 cm. The magnification is then 500. The resolution of the device is determined by the size of the spot focused by the objective. The diffraction

FIGURE 15.19 ▶ Corneal endothelial cells in an intact eye of a living bullfrog. Arrows indicate outlines of the nuclei in two of the cells. Calibration mark, 25 μm.

properties of light limit the minimum spot size to about half the wavelength of light. The optimum resolution is, therefore, about the same as in conventional microscopes.

The first biologically significant observations with the confocal microscope were those of endothelial cells on the inside of the cornea in live frogs. Such observations cannot be made with conventional microscopes because the light reflected from the front surface of the cornea masks the weak reflections from the endothelial cells. The picture of these cells shown in Fig. 15.19 was obtained by photographing the image on the oscilloscope screen. The confocal microscope is now a major observational tool in most biology laboratories. In the more recent versions of the instrument the object is scanned with moving mirrors and the image is processed by computers.

15.15.4 Fiber Optics

Fiber-optic devices are now used in a wide range of medical applications. The principle of their operation is simple. As discussed in Appendix C in connection with Snell's law, light traveling in a material of high index of refraction is totally reflected back into the material if it strikes the boundary of the material with lower refractive index at an angle greater than the critical angle θ_c. In this way, light can be confined to travel within a glass cylinder as shown in Fig. 15.21. This phenomenon has been well known since the early days of optics. However, major breakthroughs in materials technology were necessary before the phenomenon could be widely utilized.

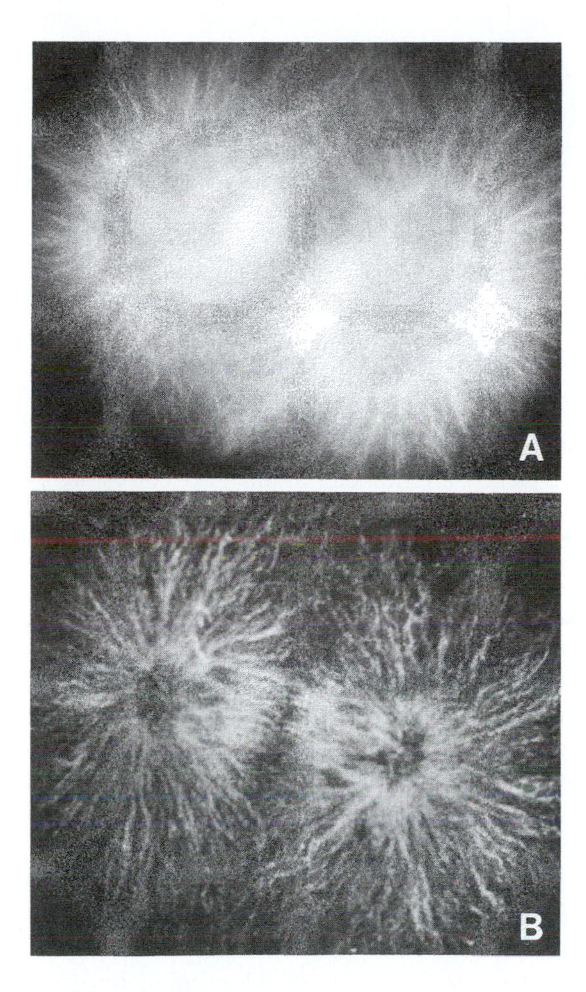

FIGURE 15.20 ▶ Microscope images of sea urchin embryos obtained with (a) a conventional microscope showing out-of-focus blur and (b) a modern confocal microscope. Part (a) from Matsumoto (1993), *Meth. Cell Biol.* **38**, p. 22. Part (b) from Wright (1989), *J. Cell. Sci.* **94**, 617–624, with permission from the Company of Biologists Ltd.

Optical fiber technology, developed in the 1960s and 1970s made it possible to manufacture low-loss, thin, highly flexible glass fibers that can carry light over long distances. A typical optical fiber is about 10 μm in diameter and is made of high purity silica glass. The fiber is coated with a cladding to increase light trapping. Such fibers can carry light over tortuously twisting paths for several kilometers without significant loss.

Fiberscopes are the simplest of the fiber-optic medical devices. They are used to visualize and examine internal organs such as the stomach, heart, and

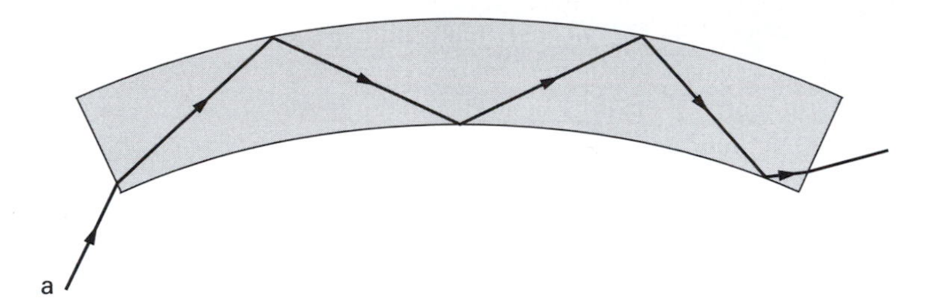

FIGURE 15.21 ▶ Light confined to travel inside a glass cylinder by total reflection.

bowels. A fiberscope consists of two bundles of optical fibers tied into one flexible unit. Each bundle is typically a millimeter in diameter consisting of about 10,000 fibers. For some applications, the bundles are thicker, up to about 1.5 cm in diameter. Depending on their use, the bundles vary in length from 0.3 to 1.2 m.

The two bundles as a unit are introduced into the body through orifices, veins, or arteries and are threaded toward the organ to be examined. Light from a high intensity source, such as a xenon arc lamp, is focused into one bundle which carries the light to the organ to be examined. Each of the fibers in the other bundle collects light reflected from a small region of the organ and carries it back to the observer. Here the light is focused into an image which can be viewed by eye or displayed on a cathode ray screen. In the usual arrangement, the illuminating bundle surrounds the light-collecting bundle.

The use of fiber-optic devices has been greatly expanded by attaching to the fiberscope remotely controlled miniature instruments to perform surgical operations without major surgical incisions. More recent applications of fiber optics include measurement of pressure in arteries, bladder, and uterus using optical sensors and laser surgery where powerful laser light is directed through one of the bundles to the tissue which is selectively destroyed.

▶ EXERCISES ▶

15-1. Compute the change in the position of the image formed by a lens with a focal length of 1.5 cm as the light source is moved from its position at 6 m from the lens to infinity.

15-2. A point source of light that is not exactly in focus produces a disk image at the retina. Assume that the image is acceptable provided the image diameter of the defocused point source is less than a. Show that the depth of field is inversely proportional to the diameter of the aperture.

15-3. Using data presented in the text, calculate the focusing power of the cornea and of the crystalline lens.

15-4. Calculate the refractive power of the cornea when it is in contact with water. The index of refraction for water is 1.33.

15-5. Calculate the focusing power of the lens in the fish eye. Assume that the lens is spherical with a diameter of 2 mm. (The indices of refraction are as in Table 15.1.) The index of refraction for water is 1.33.

15-6. Calculate the distance of the point in front of the cornea at which parallel light originating inside the reduced eye is focused.

15-7. Using the dimensions of the reduced eye (Fig. 15.5), calculate the angular resolution of the eye (use Fig. 15.6 as an aid) (a) with a single unexcited cone between points of excitation. (b) with four unexcited cones between areas of excitation.

15-8. Calculate the distance from which a person with good vision can see the whites of another person's eyes. Use data in the text and assume the size of the eye is 1 cm.

15-9. Calculate the size of the retinal image of a 10-cm leaf from a distance of 500 m.

Chapter 16

Atomic and Nuclear Physics

Modern atomic and nuclear physics is among the most impressive scientific achievements of this century. There is hardly an area of science or technology that does not draw on the concepts and techniques developed in this field. Both the theories and techniques of atomic and nuclear physics have played an important role in the life sciences. The theories provided a solid foundation for understanding the structure and interaction of organic molecules, and the techniques provided many tools for both experimental and clinical work. Contributions from this field have been so numerous and influential that it is impossible to do them justice in a single chapter. Of necessity, therefore, our discussion will be restricted to a survey of the subject. We will present a brief description of the atom and the nucleus, which will lead into a discussion of the applications of atomic and nuclear physics to the life sciences.

16.1 The Atom

By 1912, through the work of J. J. Thompson, E. Rutherford, and their colleagues, a number of important facts had been discovered about atoms which make up matter. It was found that atoms contain small negatively charged electrons and relatively heavier positively charged protons. The proton is about 2000 times heavier than the electron, but the magnitude of the charge on the two is the same. There are as many positively charged protons in an atom as negatively charged electrons. The atom as a whole is, therefore, electrically neutral. The identity of an atom is determined by the number of protons it has. For example, hydrogen has 1 proton, carbon has 6 protons, silver has 47 protons. Through a series of ingenious experiments, Rutherford showed that most of the atomic mass is concentrated in the nucleus consisting of protons and that the electrons are somehow situated outside the nucleus. It was subsequently discovered that the nucleus also contains another particle, the neutron, which has approximately the same mass as the proton but is electrically neutral.

Although the nucleus contains most of the atomic mass, it occupies only a small part of the total atomic volume. The diameter of the whole atom is on the order of 10^{-8} cm, but the diameter of the nucleus is only about 10^{-13} cm. The configuration of the electrons around the nucleus was not known at that time.

In 1913, the Danish physicist Niels Bohr proposed a model for the atom which explained many observations that were puzzling scientists at that time. When Bohr first became acquainted with atomic physics, the subject was in a state of confusion. A number of theories had been proposed for the structure of the atom, but none explained satisfactorily the existing experimental results. The most surprising observed property of atoms was the light emitted by them.[1] When an element is put into a flame, it emits light at sharply defined wavelengths, called *spectral lines*. Each element emits its own characteristic spectrum of light. This is in contrast to a glowing filament in a light bulb, for example, which emits light over a continuous range of wavelengths.

Prior to Bohr, scientists could not explain why these colors were emitted by atoms. Bohr's model of the atom explained the reason for the sharp spectra. Bohr started with the model of the atom as proposed by Rutherford. At the center of the atom is the positive nucleus made up of protons (and neutrons). The electrons orbit around the nucleus much as the planets orbit around the sun. They are maintained in orbit by the electrostatic attraction of the nucleus. And here is the major feature of the Bohr model: So that the model would explain the emission of spectral lines, Bohr had to postulate that the electrons are restricted to distinct orbits around the nucleus. In other words, the electrons can be found only in certain allowed orbits. Bohr was able to calculate the radii of these allowed orbits and show that the spectral lines are emitted as a consequence of the orbital restrictions. Bohr's calculations are found in most elementary physics texts.

The orbital restrictions are most easily illustrated with the simplest atom, hydrogen, which has a single-proton nucleus and one electron orbiting around it (Fig. 16.1). Unless energy is added to the atom, the electron is found in the allowed orbit closest to the nucleus. If energy is added to the atom, the

[1] In atomic physics, the word *light* is not restricted solely to the visible part of the electromagnetic spectrum. Radiation at shorter wavelength (ultraviolet) and longer wavelength (infrared) is also often referred to as light.

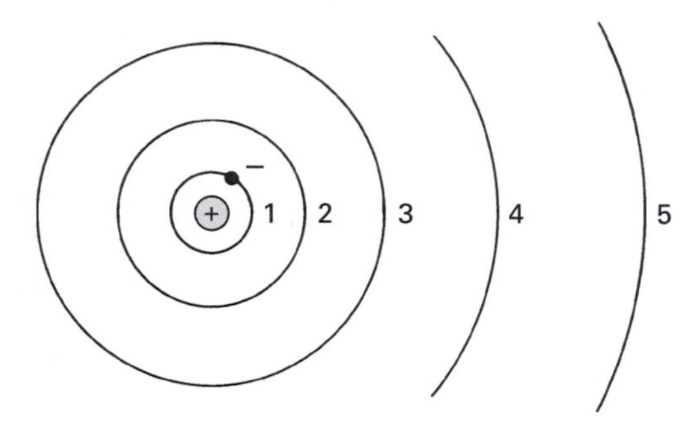

FIGURE 16.1 ▶ Bohr model for the hydrogen atom. The electron orbits about the nucleus and can occupy only discrete orbits with radii 1, 2, 3, and so on.

electron may "jump" to one of the higher allowed orbits farther away from the nucleus, but the electron can never occupy the regions between the allowed orbits.

The Bohr model was very successful in explaining many of the experimental observations for the simple hydrogen atom. But to describe the behavior of atoms with more than one electron, it was necessary to impose an additional restriction on the structure of the atom: The number of electrons in a given orbit cannot be greater than $2n^2$, where n is the order of the orbit from the nucleus. Thus, the maximum number of electrons in the first allowed orbit is $2 \times (1)^2 = 2$; in the second allowed orbit, it is $2 \times (2)^2 = 8$; in the third orbit, it is $2 \times (3)^2 = 18$, and so on.

The atoms are found to be constructed in accordance with these restrictions. Helium has two electrons, and, therefore, its first orbit is filled. Lithium has three electrons, two of which fill the first orbit; the third electron, therefore, must be in the second orbit. This simple sequence is not completely applicable to the very complex atoms, but basically this is the way the elements are constructed.

A specific amount of energy is associated with each allowed orbital configuration of the electron. Therefore, instead of speaking of the electron as being in a certain orbit, we can refer to it as having a corresponding amount of energy. Each of these allowed values of energy is called an *energy level.* An energy level diagram for an atom is shown in Fig. 16.2. Note that every element has its own characteristic energy level structure. The electrons in the atom can occupy only specific energy states; that is, in a given atom the electron can have an energy E_1, E_2, E_3, and so on, but cannot have an energy between these two values. This is a direct consequence of the restrictions on the allowed electron orbital configurations.

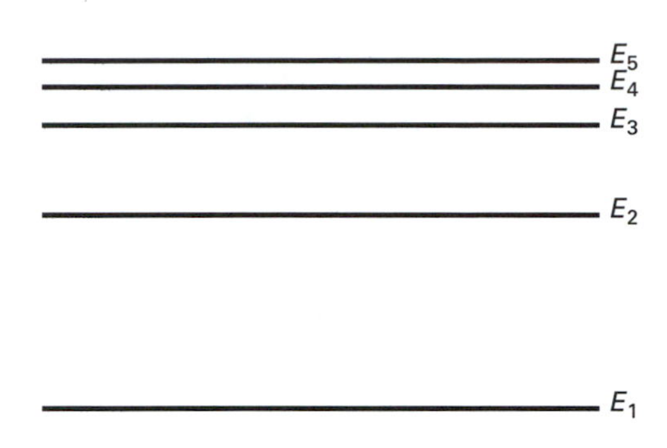

FIGURE 16.2 ▶ Energy levels for an atom.

The lowest energy level that an electron can occupy is called the *ground state*. This state is associated with the orbital configuration closest to the nucleus. The higher allowed energy levels, called *excited states*, are associated with larger orbits and different orbital shapes. Normally the electron occupies the lowest energy level but it can be excited into a higher energy state by adding energy to the atom.

An atom can be excited from a lower to a higher energy state in a number of different ways. The two most common methods of excitation are electron impact and absorption of electromagnetic radiation. Excitation by electron impact occurs most frequently in a gas discharge. If a current is passed through a gas of atoms, the colliding electron is slowed down and the electron in the atom is promoted to a higher energy configuration. When the excited atoms fall back into the lower energy states, the excess energy is given off as electromagnetic radiation. Each atom releases its excess energy in a single photon. Therefore, the energy of the photon is simply the difference between the energies of the initial state E_i and the final state E_f of the atom. The frequency f of the emitted radiation is given by

$$f = \frac{\text{Energy of photon}}{\text{Planck constant}} = \frac{E_i - E_f}{h} \qquad (16.1)$$

Transition between each pair of energy levels results in the emission of light at a specific frequency. Therefore, a group of highly excited atoms of a given element emit light at a number of well-defined frequencies which constitute the optical spectrum for that element.

An atom in a given energy level can also be excited to a higher level by light at a specific frequency. The frequency must be such that each photon has just the right amount of energy to promote the atom to one of its higher allowed energy states. Atoms, therefore, absorb light only at specific frequencies, given by Eq. 16.1. Light at other frequencies is not absorbed. If a beam of white light (containing all the frequencies) is passed through a group of

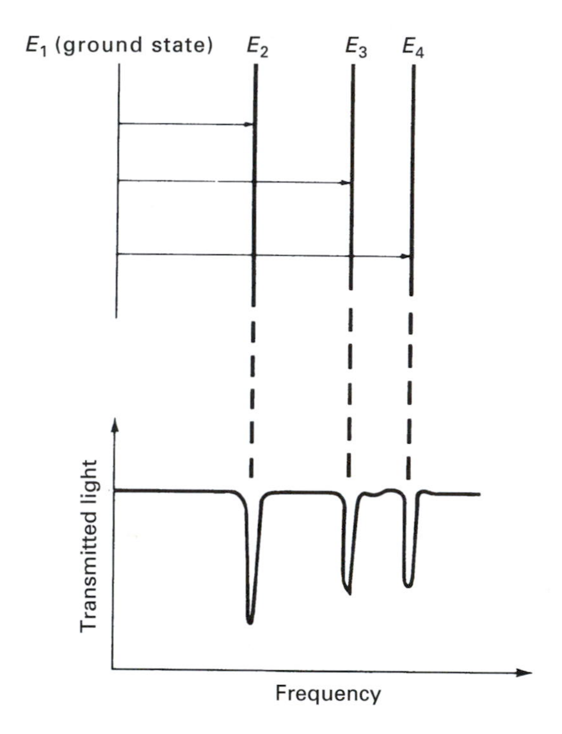

FIGURE 16.3 ▶ The absorption spectrum.

atoms of a given species, the spectrum of the transmitted light shows gaps corresponding to the absorption of the specific frequencies by the atoms. This is called the *absorption spectrum* of the atom. In their undisturbed state, most of the atoms are in the ground state. The absorption spectrum, therefore, usually contains only lines associated with transitions from the ground state to higher allowed states (Fig. 16.3).

Optical spectra are produced by the outer electrons of the atom. The inner electrons, those closer to the nucleus, are bound more tightly and are consequently more difficult to excite. However, in a highly energetic collision with another particle, an inner electron may be excited. When in such an excited atom an electron returns to the inner orbit, the excess energy is again released as a quantum of electromagnetic radiation. Because the binding energy here is about a thousand times greater than for the outer electrons, the frequency of the emitted radiation is correspondingly higher. Electromagnetic radiation in this frequency range is called *X-rays*.

The Bohr model also explained qualitatively the formation of chemical bonds. The formation of chemical compounds and matter in bulk is due to

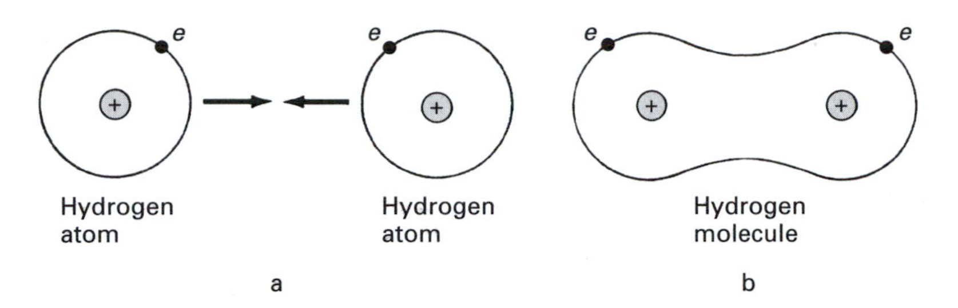

FIGURE 16.4 ▶ A schematic representation for the formation of a hydrogen molecule. (a) Two separate hydrogen atoms. (b) When the two atoms are close together, the electrons share each other's orbit, which results in the binding of the two atoms into a molecule.

the distribution of electrons in the atomic orbits. When an orbit is not filled to capacity (which is the case for most atoms), the electrons of one atom can partially occupy the orbit of another. This sharing of orbits draws the atoms together and produces bonding between atoms. As an example, we show in Fig. 16.4 the formation of a hydrogen molecule from two hydrogen atoms. In the orbit of each of the hydrogen atoms there is room for another electron. A completely filled orbit is the most stable configuration; therefore, when two hydrogen atoms are close together, they share each other's electrons, and, in this way, the orbit of each atom is completely filled part of the time. This shared orbit can be pictured as a rubber band pulling the two atoms together. Therefore, the sharing of the electrons binds the atoms into a molecule. While the sharing of electrons pulls the atoms together, the coulomb repulsion of the nuclei tends to keep them apart. The equilibrium separation between atoms in a molecule is determined by these two counter forces. In a similar way, more complex molecules, and ultimately bulk matter, are formed.

Atoms with completely filled orbits (these are atoms of the so-called *noble gases*—helium, neon, argon, krypton, and xenon) cannot share electrons with other elements and are, therefore, chemically most inert.

Molecules also have characteristic spectra both in emission and in absorption. Because molecules are more complicated than atoms, their spectra are correspondingly more complex. In addition to the electronic configuration, these spectra also depend on the motion of the nuclei. Still the spectra can be interpreted and are unique for each type of molecule.

16.2 Spectroscopy

The absorption and emission spectra of atoms and molecules are unique for each species. They can serve as fingerprints in identifying atoms and molecules in various substances. Spectroscopic techniques were first used in basic experiments with atoms and molecules, but they were soon adopted in many other areas, including the life sciences.

In biochemistry, spectroscopy is used to identify the products of complex chemical reactions. In medicine, spectroscopy is used routinely to determine the concentration of certain atoms and molecules in the body. From a spectroscopic analysis of urine, for example, one can determine the level of mercury in the body. Blood-sugar level is measured by first producing a chemical reaction in the blood sample which results in a colored product. The concentration of this colored product, which is proportional to the blood-sugar level, is then measured by absorption spectroscopy.

The basic principles of spectroscopy are simple. In emission spectroscopy the sample under investigation is excited by an electric current or a flame. The emitted light is then examined and identified. In absorption spectroscopy, the sample is placed in the path of a beam of white light. Examination of the transmitted light reveals the missing wavelengths which identify the components in the substance. Both the absorption and the emission spectra can provide information also about the concentration of the various components in the substance. In the case of emission, the intensity of the emitted light in the spectrum is proportional to the number of atoms or molecules of the given species. In absorption spectroscopy, the amount of absorption can be related to the concentration. The instrument used to analyze the spectra is called a *spectrometer*. This device records the intensity of light as a function of wavelength.

A spectrometer, in its simplest form, consists of a focusing system, a prism, and a detector for light (see Fig. 16.5). The focusing system forms a parallel beam of light which falls on the prism. The prism, which can be rotated, breaks up the beam into its component wavelengths. At this point, the fanned-out spectrum can be photographed and identified. Usually, however, the spectrum is detected a small section at a time. This is accomplished by the narrow exit slit which intercepts only a portion of the spectrum. As the prism is rotated, the whole spectrum is swept sequentially past the slit. The position of the prism is calibrated to correspond with the wavelength impinging on the slit. The light that passes through the slit is detected by a photodetector which produces an electrical signal proportional to the light intensity. The intensity of the signal as a function of wavelength can be displayed on a chart recorder.

Spectrometers used in routine clinical work are automated and can be operated by relatively unskilled personnel. The identification and interpretation of the spectra produced by less well-known molecules, however, require considerable training and skill. In addition to identifying the molecule, such spectra also yield information about the molecular structure. The use of spectrometers is further explored in Exercise 6-1.

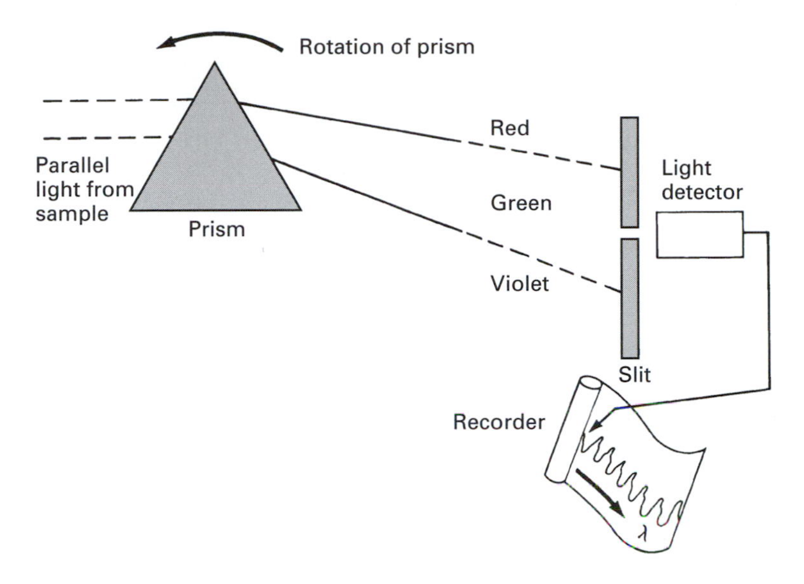

FIGURE 16.5 ▶ The measurement of spectra.

16.3 Quantum Mechanics

Although the Bohr model explained many observations, from the very beginning the theory appeared contrived. Certainly the concept of stable allowed orbits with a specific number of electrons seemed arbitrary. The model, however, was a daring step in a new direction that eventually led to the development of quantum mechanics.

In the quantum mechanical description of the atom, it is not possible to assign exact orbits or trajectories to electrons. Electrons possess wavelike properties and behave as clouds of specific shape around the nucleus. The artificial postulates in Bohr's theory are a natural consequence of the quantum mechanical approach to the atom. Furthermore, quantum mechanics explains many phenomena outside the scope of the Bohr model. The shape of simple molecules, for example, can be shown to be the direct consequence of the interaction between the electron configurations in the component atoms.

The concept that particles may exhibit wavelike properties was introduced in 1924 by Louis de Broglie. This suggestion grew out of an analogy with light which was then known to have both wave- and particlelike properties. De Broglie suggested by analogy that particles may exhibit wavelike properties. He showed that the wavelength λ of the matter waves would be

$$\lambda = \frac{h}{mv} \tag{16.2}$$

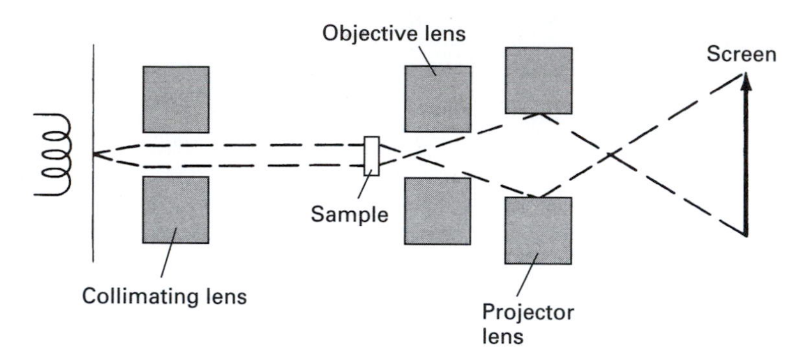

FIGURE 16.6 ▶ The electron microscope.

where m and v are the mass and velocity of the particle and h is the Planck constant.

In 1925, de Broglie's hypothesis was confirmed by experiments which showed that electrons passing through crystals form wavelike diffraction patterns with a configuration corresponding to a wavelength given by Eq. 16.2.

16.4 Electron Microscope

In Chapter 15, we pointed out that the size of the smallest object observable by a microscope is about half the wavelength of the illuminating radiation. In light microscopes, this limits the resolution to about 200 nm (2000 Å). Because of the wave properties of electrons, it is possible to construct microscopes with a resolution nearly 1000 times smaller than this value.

It is relatively easy to accelerate electrons in an evacuated chamber to high velocities so that their wavelength is less than 10^{-10} m (1 Å). Furthermore, the direction of motion of the electrons can be altered by electric and magnetic fields. Thus, suitably shaped fields can act as lenses for the electrons. The short wavelength of electrons coupled with the possibility of focusing them has led to the development of electron microscopes that can observe objects 1000 times smaller than are visible with light microscopes. The basic construction of an electron microscope is shown in Fig. 16.6. The similarities between the electron and the light microscope are evident: Both have the same basic configuration of two lenses which produce two-stage magnification. Electrons are emitted from a heated filament and are then accelerated and collimated into a beam. The beam passes through the thin sample under examination which diffracts the electrons in much the same way as light is diffracted in an optical microscope. But because of their short wavelength,

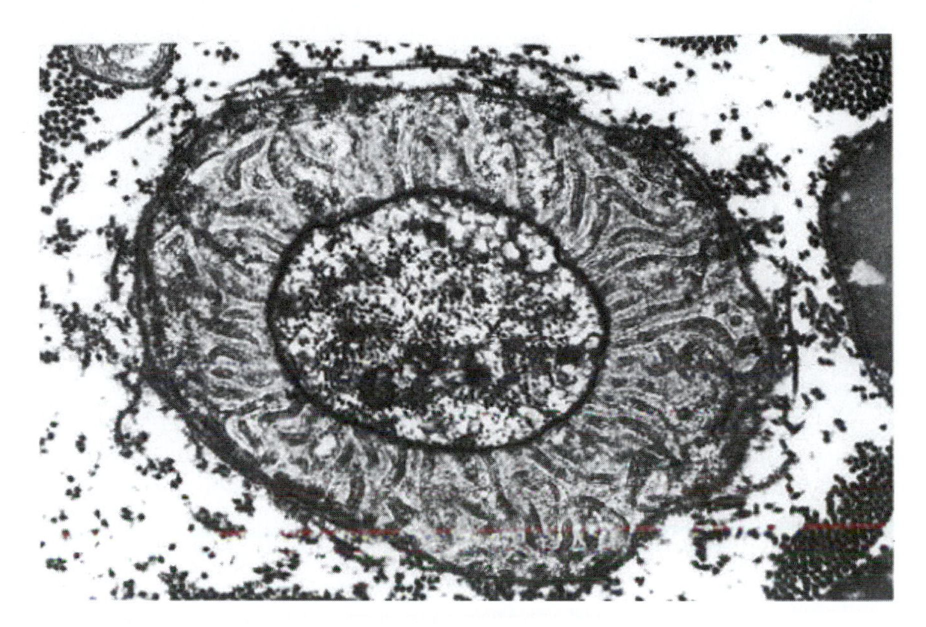

FIGURE 16.7 ▶ Electron micrograph of an individual axon in the peripheral nerve of a mouse. The cross section of the axon at the level of the node of Ranvier is about 2.5 μm in width. Surrounding the axon is a differentiated region of the myelin sheath. (Photograph courtesy of Professor Dan Kirschner, Biology Department, Boston College, and Dr. Bela Kosaras, Primate Center, Southborough, MA.)

the electrons are influenced by much smaller structures within the sample. The transmitted electrons are focussed into a real image by the objective lens. This image is then further magnified by the projector lens, which projects the final image onto film or a fluorescent screen. Although it is possible to produce electrons with a wavelength much less than 10^{-10} m (1 Å), the theoretical optimum resolution implied by such short wavelengths has not yet been realized. At present, the best resolution of electron microscopes is about 5×10^{-10} m (5 Å).

Because electrons are scattered by air, the microscope must be contained in an evacuated chamber. Furthermore, the samples under examination must be dry and thin. These conditions, of course, present some limitations in the study of biological materials. The samples have to be specially prepared for electron microscopic examination. They must be dry, thin, and in some cases coated. Nevertheless, electron microscopes have yielded beautiful pictures showing details in cell structure, biological processes, and recently even large molecules such as DNA in the process of replication (see Fig. 16.7).

16.5 X-rays

In 1895, Wilhelm Conrad Roentgen announced his discovery of X-rays. He had found that when high-energy electrons hit a material such as glass, the material emitted radiation that penetrated objects which are opaque to light. He called this radiation X-rays. It was shown subsequently that X-rays are short-wavelength, electromagnetic radiation emitted by highly excited atoms. Roentgen showed that X-rays could expose film and produce images of objects in opaque containers. Such pictures are possible if the container transmits X-rays more readily than the object inside. A film exposed by the X-rays shows the shadow cast by the object.

Within three weeks of Roentgen's announcement, two French physicians, Oudin and Barthélemy, obtained X-rays of bones in a hand. Since then, X-rays have become one of the most important diagnostic tools in medicine. With current techniques, it is even possible to view internal body organs that are quite transparent to X-rays. This is done by injecting into the organ a fluid opaque to X-rays. The walls of the organ then show up clearly by contrast.

X-rays have also provided valuable information about the structure of biologically important molecules. The technique used here is called *crystallography*. The wavelength of X-rays is on the order of 10^{-10}m, about the same as the distance between atoms in a molecule or crystal. Therefore, if a beam of X-rays is passed through a crystal, the transmitted rays produce a diffraction pattern that contains information about the structure and composition of the crystal. The diffraction pattern consists of regions of high and low X-ray intensity which when photographed show spots of varying brightness (Fig. 16.8).

Diffraction studies are most successfully done with molecules that can be formed into a regular periodic crystalline array. Many biological molecules

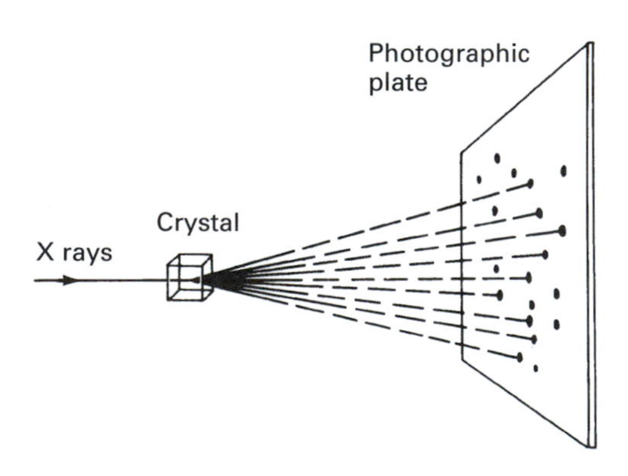

FIGURE 16.8 ▶ Arrangement for detecting diffraction of X-rays by a crystal.

can in fact be crystallized under the proper conditions. It should be noted, however, that the diffraction pattern is not a unique, unambiguous picture of the molecules in the crystal. The pattern is a mapping of the collective effect of the arrayed molecules on the X-rays that pass through the crystal. The structure of the individual molecule must be deduced from the indirect evidence provided by the diffraction pattern.

If the crystal has a simple structure—such as sodium chloride, for example—the X-ray diffraction pattern is also simple and relatively easy to interpret. Complicated crystals, however, such as those synthesized from organic molecules, produce very complex diffraction patterns. But, even in this case, it is possible to obtain some information about the structure of the molecules forming the crystal (for details, see [16-1]). To resolve the three-dimensional features of the molecules, diffraction patterns must be formed from thousands of different angles. The patterns are then analyzed, with the aid of a computer. These types of studies provided critical information for the determination of the structure for penicillin, vitamin B_{12}, DNA, and many other biologically important molecules.

16.6 X-ray Computerized Tomography

The usual X-ray picture does not provide depth information. The image represents the total attenuation as the X-ray beam passes through the object in its path. For example, a conventional X-ray of the lung may reveal the existence of a tumor, but it will not show how deep in the lung the tumor is located. Several *tomographic techniques* (CT scans) have been developed to produce slice-images within the body which provide depth information. (Tomography is from the Greek word *tomos* meaning section.) Presently the most commonly used of these is X-ray computerized tomography (CT scan) developed in the 1960s. The basic principle of the technique in its simplest form is illustrated in Fig. 16.9a and b. A thin beam of X-rays passes through the plane we want to visualize and is detected by a diametrically opposing detector. For a given angle with respect to the object (in this case the head), the X-ray source-detector combination is moved laterally scanning the region of interest as shown by the arrow in Fig. 16.9a. At each position, the detected signal carries integrated information about X-ray transmission properties of the full path in this case A–B. The angle is then changed by a small amount (about 1°) and the process is repeated full circle around the object. As indicated in Fig. 16.9b, by rotating the source-detector combination, information can be obtained about the intersection points of the X-ray beams.

In Fig. 16.9b, we show schematically the scanning beam at two angles with two lateral positions at each angle. While at each position, the detected

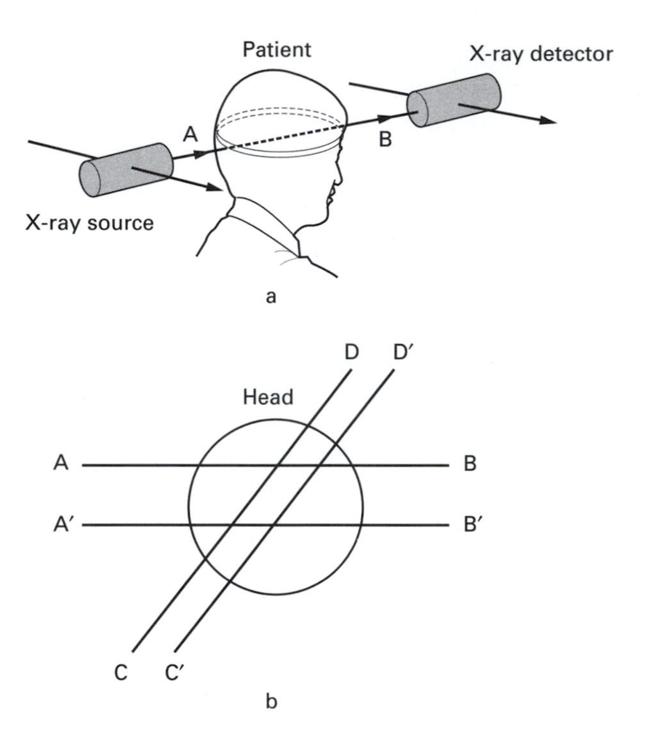

FIGURE 16.9 ▶ (a) Basic principle of computed axial tomography. (b) Rotation of the source-detector combination provides information about the X-ray transmission properties of each point within the plane of the object to be studied.

signal carries integrated information about the full path, two paths that intersect contain common information about the one point of intersection. In the figure, four such points are shown at the intersection of the beams $A–B$, $A'–B'$, $C–D$, and $C'–D'$. The multiple images obtained by translation and rotation contain information about the X-ray transmission properties of each point within the plane of the object to be studied. These signals are stored and by a rather complex computer analysis a point by point image is constructed of the thin slice scanned within the body.

The visualized slices within the body obtained in this way are typically about 2 mm thick. In the more recent versions of the instrument, a fan rather than a beam of X-rays scans the object, and an array of multiple detectors is used to record the signal. Data acquisition is speeded up in this way yielding an image in a few seconds.

16.7 The Nucleus

Although all the atoms of a given element have the same number of protons in their nucleus, the number of neutrons may vary. Atoms with the same number of protons but different number of neutrons are called *isotopes*. All the nuclei of the oxygen atom, for example, contain 8 protons but the number of neutrons in the nucleus may be 8, 9, or 10. These are the isotopes of oxygen. They are designated as ${}^{16}_{8}O$, ${}^{17}_{8}O$, and ${}^{18}_{8}O$. This is a general type of nuclear symbolism in which the subscript to the chemical symbol of the element is the number of protons in the nucleus and the superscript is the sum of the number of protons and neutrons. The number of neutrons often determines the stability of the nucleus.

The nuclei of most naturally occurring atoms are stable. They do not change when left alone. There are, however, many unstable nuclei which undergo transformations accompanied by the emission of energetic radiation. It has been found that the emanations from these radioactive nuclei fall into three categories: (1) alpha (α) particles, which are high-speed helium nuclei consisting of two protons and two neutrons; (2) beta (β) particles, which are very high-speed electrons; and (3) gamma (γ) rays, which are highly energetic photons.

The radioactive nucleus of a given element does not emit all three radiations simultaneously. Some nuclei emit alpha particles, others emit beta particles, and the emission of gamma rays may accompany either event.

Radioactivity is associated with the transmutation of the nucleus from one element to another. Thus, for example, when radium emits an alpha particle, the nucleus is transformed into the element radon. The details of the process are discussed in most physics texts (see [16-2]).

The decay or transmutation of a given radioactive nucleus is a random event. Some nuclei decay sooner; others decay later. If, however, we deal with a large number of radioactive nuclei, it is possible, by using the laws of probability, to predict accurately the decay rate for the aggregate. This decay rate is characterized by the half-life, which is the time interval for half the original nuclei to undergo transmutation.

There is a great variation in the half-life of radioactive elements. Some decay very quickly and have a half-life of only a few microseconds or less. Others decay slowly with a half-life of many thousands of years. Only the very long-lived radioactive elements occur naturally in the Earth's crust. One of these, for example, is the uranium isotope ${}^{238}_{92}U$, which has a half-life of 4.51×10^9 years. The short-lived radioactive isotopes can be produced in accelerators by bombarding certain stable elements with high-energy particles. Naturally occurring phosphorus, for example, has 15 protons and 16 neutrons in its nucleus (${}^{31}_{15}P$). The radioactive phosphorus isotope ${}^{32}_{15}P$ with 17 neutrons can be produced by bombarding sulfur with neutrons. The reaction is

$$^{32}_{16}\text{S} + \text{neutron} \rightarrow {}^{32}_{15}\text{P} + \text{proton}$$

This radioactive phosphorus has a half-life of 14.3 days. Radioactive isotopes of other elements can be produced in a similar way. Many of these isotopes have been very useful in biological and clinical work.

16.8 Radiation Therapy

The photons of X-rays and gamma-rays and the particles emitted by radioactive nuclei all have energies far greater than the energies that bind electrons to atoms and molecules. As a result, when such radiation penetrates into biological materials, it can rip off electrons from the biological molecules and produce substantial alterations in their structure. The ionized molecule may break up, or it may chemically combine with another molecule to form an undesirable complex. If the damaged molecule is an important component of a cell, the whole cell may die. Water molecules in the tissue are also broken up by the radiation into reactive fragments (H + OH). These fragments combine with the biological molecules and alter them in a detrimental way. In addition, radiation passing through tissue may simply give up its energy and heat the tissue to a dangerously high temperature. A large dose of radiation may damage so many cells that the whole organism dies. Smaller but still dangerous doses may produce irreversible changes such as mutations, sterility, and cancer.

In controlled doses, however, radiation can be used therapeutically. In the treatment of certain types of cancer, an ampul containing radioactive material such as radium or cobalt 60 is implanted near the cancerous growth. By careful placement of the radioactive material and by controlling the dose, the hope is to destroy the cancer without greatly damaging the healthy tissue. Unfortunately some damage to healthy tissue is unavoidable. As a result, this treatment is often accompanied by the symptoms of radiation sickness (diarrhea, nausea, loss of hair, loss of appetite, and so on). If long-lived isotopes are used in the therapy, the material must be removed after a prescribed period. Short-lived isotopes, such as gold 198 with a half-life of about 3 days, decay quickly enough so that they do not need to be removed after treatment.

Certain elements introduced into the body by injection or by mouth tend to concentrate in specific organs. This phenomenon is used to advantage in radiation therapy. The radioactive isotope phosphorus 32 (half-life, 14.3 days) mentioned earlier accumulates in the bone marrow. Iodine 131 (half-life, 8 days) accumulates in the thyroid and is given for the treatment of hyperthyroidism.

An externally applied beam of gamma rays or X-rays can also be used to destroy cancerous tumors. The advantage here is that the treatment is administered without surgery. The effect of radiation on the healthy tissue can be reduced by frequently altering the direction of the beam passing through the body. The tumor is always in the path of the beam, but the dosage received by a given section of healthy tissue is reduced.

16.9 Food Preservation by Radiation

Without some attempt at preservation, all foods decay rather quickly. Within days and often within hours, many foods spoil to a point where they become inedible. The decay is usually caused by microorganisms and enzymes that decompose the organic molecules of the food.

Over the years, a number of techniques have been developed to retard spoilage. Keeping the food in a cold environment reduces the rate of activity for both the enzymes and the microorganisms. Dehydration of food achieves the same goal. Heating the food for a certain period of time destroys many microorganisms and again retards decay. This is the principle of *pasteurization*. These methods of retarding spoilage are all at least 100 years old. There is now a new technique of preserving food by irradiation.

High-energy radiation passing through the food destroys microorganisms that cause decay. Radiation is also effective in destroying small insects that attack stored foods. This is especially important for wheat and other grains which at present are often fumigated before shipping or storage. Chemical fumigation kills the insects but not their eggs. When the eggs hatch, the new insects may destroy a considerable fraction of the grain. Radiation kills both the insects and the eggs.

Gamma rays are used most frequently in food preservation. They have a great penetrating power and are produced by relatively inexpensive isotopes such as cobalt 60 and cesium 137. High-speed electrons produced by accelerators have also been used to sterilize food. Electrons do not have the penetrating power of gamma rays, but they can be aimed better and can be turned off when not in use.

In the United States and in many other countries, there are now a number of facilities for irradiating food. In the usual arrangement, the food on a conveyor passes by the radioactive source, where it receives a controlled dose of radiation. The source must be carefully shielded to protect the operator. This problem is relatively simple to solve, and at present the technical problems seem to be well in hand. One plant for irradiating food, in Gloucester, Massachusetts, initially built by the Atomic Energy Commission, has been operating successfully since 1964. It can process 1000 lb of fish per hour.

There is no doubt that irradiation retards spoilage of food. Irradiated strawberries, for example, remain fresh for about 15 days after they have been picked whereas strawberries that have not been treated begin to decay after about 10 days. Irradiated unfrozen fish also lasts a week or two longer. Tests have shown that the taste, nutritional value, and appearance of the food remain acceptable. The important question is the safety of the procedure. Irradiation at the levels used in the treatment does not make the food radioactive. There is, however, the possibility that the changes induced by radiation may make the food harmful. Over the past three decades, there have been many test programs both with animals and with human volunteers to ascertain the safety of food irradiation. At this point, the technique has been judged safe and is in commercial use (see Exercise 10-3).

16.10 Isotopic Tracers

Most elements have isotopes differing from each other by the number of neutrons in their nuclei. The isotopes of a given element are chemically identical—that is, they participate in the same chemical reactions—but they can be distinguished from each other because their nuclei are different. One difference is, of course, in their mass. This property alone can be used to separate one isotope from another. A mass spectrometer is one of the devices that can perform this task. Another way to distinguish isotopes is by their radioactivity. Many elements have isotopes that are radioactive. These isotopes are easily identified by their activity. In either case, isotopes can be used to trace the various steps in chemical reactions and in metabolic processes. Tracer techniques have been useful also in the clinical diagnoses of certain disorders.

Basically the technique consists of introducing a rare isotope into the process and then following the course of the isotope with appropriate detection techniques. We will illustrate this technique with a few examples. Nitrogen is one of the atoms in the amino acids that compose the protein molecules. In nature, nitrogen is composed primarily of the isotope ^{14}N. Only 0.36% of natural nitrogen is in the form of the nonradioactive isotope ^{15}N. Ordinarily the amino acids reflect the natural composition of nitrogen.

It is possible to synthesize amino acids in a laboratory. If the synthesis is done with pure ^{15}N, the amino acids are distinctly marked. The amino acid glycine produced in this way is introduced into the body of a subject where it is incorporated into the hemoglobin of the blood. Periodic sampling of the blood measures the number of blood cells containing the originally introduced glycine. Such experiments have shown that the average lifetime of a red blood cell is about four months.

Radioactive isotopes can be traced more easily and in smaller quantities than the isotopes that are not radioactive. Therefore, in reactions with elements that have radioactive isotopes, radioactive tracer techniques are preferred. Since the 1950s, when radioactive isotopes first became widely available, hundreds of important experiments have been conducted in this field.

An example of this technique is the use of radioactive phosphorus in the study of nucleic acids. The element phosphorus is an important component of the nucleic acids DNA and RNA. Naturally occurring phosphorus is all in the form ^{31}P, and, of course, this is the isotope normally found in the nucleic acids. However, as discussed earlier, by bombarding sulphur 32 with neutrons, it is possible to produce the radioactive phosphorus ^{32}P which has a half-life of 14.3 days. If the ^{32}P isotope is introduced into the cell, the nucleic acids synthesized in the cell incorporate this isotope into their structure. The nucleic acids are then removed from the cell and their radioactivity is measured. From these measurements it is possible to calculate the rate at which nucleic acids are manufactured by the cell. These measurements, among others, provided evidence for the roles of DNA and RNA in cell functions.

Radioactive tracers have been useful also in clinical measurements. In one technique, the radioactive isotope of chromium is used to detect internal hemorrhage. This isotope is taken up by the blood cells, which then become radioactive. The radioactivity is, of course, kept well below the danger level. If the circulation is normal, the radioactivity is distributed uniformly throughout the body. A pronounced increase in radioactivity in some region indicates a hemorrhage at that point.

16.11 Magnetic Resonance Imaging

Images of the shapes of internal organs obtained with computerized X-ray tomography are excellent. However, X-rays, do not provide information about the internal structure of tissue. CT scans may therefore, miss changes in tissue structure and pathological alteration inside internal organs. *Magnetic resonance imaging* (*MRI*), introduced in the early 1980s, is the most recent addition to medical imaging techniques. This technique utilizes the magnetic properties of the nucleus to provide images of internal body organs with detailed information about soft-tissue structure.

The imaging techniques we have discussed so far (X-ray and ultrasound) are in principle relatively simple. They utilize reflected or transmitted energy to visualize internal structures. Magnetic resonance imaging is more complex. It utilizes the principles of *nuclear magnetic resonance* (*NMR*) developed in the 1940s. A detailed description of MRI is beyond the scope of this text, but the principles are relatively simple to explain. A discussion of MRI begins with an introduction to the principles of nuclear magnetic resonance.

16.11.1 Nuclear Magnetic Resonance

Protons and neutrons which are the constituents of atomic nuclei possess the quantum mechanical property of spin which has magnitude and direction. We can imagine these particles as if they were small spinning tops. As a result of spin, the nuclear particles act as small bar magnets. Inside the nucleus, these small magnets associated with the nucleons (protons and neutrons) line up so as to cancel each other's magnetic fields. However, if the number of nucleons is odd, the cancellation is not complete, and the nucleus possesses a net magnetic moment. Therefore, nuclei with an odd number of nucleons behave as tiny magnets. Hydrogen, which has a nucleus consisting of a single proton, does, of course, have a nuclear magnetic moment. The human body is made of mostly water and other hydrogen-containing molecules. Therefore, MRI images of structures within the body can be most effectively produced using the magnetic properties of the hydrogen nucleus. Our discussion will be restricted to the nuclear magnetic properties of hydrogen.

Normally, the little nuclear magnets in bulk material are randomized in space as is shown in Fig. 16.10a, and the material does not possess a net magnetic moment ($M = 0$). The nuclear magnets are represented as small arrows. However, the situation is altered in the presence of an external magnetic field. When an external magnetic field is applied to a material possessing nuclear magnetic moments, the tiny nuclear magnets line up either parallel or antiparallel with the magnetic field as shown in Fig. 16.10b. The direction of the external magnetic field is usually designated as the z-axis. As shown in the figure, the x-y plane is orthogonal to the z-axis. Because the nuclear magnets parallel to the field ($+z$) have a somewhat lower energy than those that are antiparallel ($-z$), more of the nuclei are in the parallel state than in the antiparallel state. In an external magnetic field, the assembly of parallel/antiparallel nuclear spins as a whole has a net magnetic moment M that behaves as a magnet pointing in the direction of the magnetic field.

The energy spacing ΔE_m between the parallel and antiparallel alignments is

$$\Delta E_m = \frac{\gamma h B}{2\pi} \tag{16.3}$$

Here B is the externally applied magnetic field, h is the Planck constant as defined earlier, and γ is called *gyromagnetic ratio* which is a property of a given nucleus. Typically the strength of magnetic fields used in MRI is about 2 tesla (T). (By comparison, the strength of the magnetic field of the Earth is on the order of 10^{-4} T.)

The discrete energy spacing ΔE_m, between the two states shown in Fig. 16.10b, makes this a resonant system. The frequency corresponding to the energy difference between the two states is called the *Larmor frequency* and

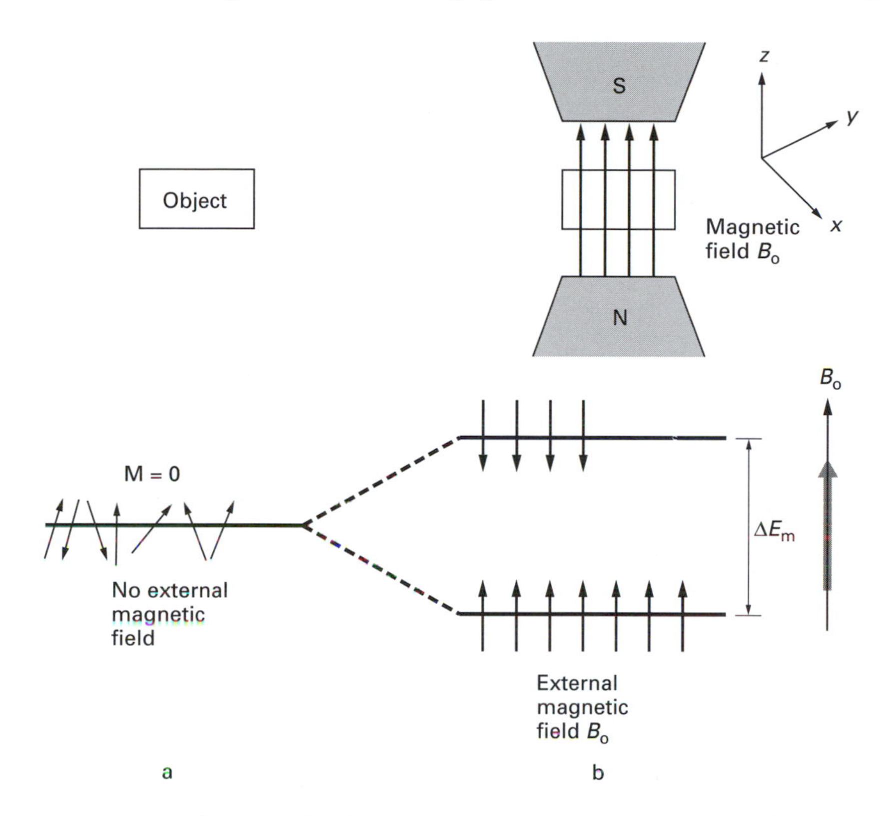

FIGURE 16.10 ▶ (a) In the absence of an external magnetic field, nuclear spins are randomized. (b) When an external magnetic field is applied to a material possessing nuclear magnetic moments, the tiny nuclear magnets line up either parallel or antiparallel with the magnetic field. The parallel configuration is at a lower energy.

in accord with Eq. 16.1 is given by

$$f_L = \frac{\Delta E_m}{h} = \frac{\gamma B}{2\pi} \tag{16.4}$$

The gyromagnetic ratio γ for a proton is 2.68×10^8 $T^{-1}sec^{-1}$. Magnetic fields used in MRI are typically in the range 1 to 4 T. The corresponding Larmor frequencies are about 43 to 170 MHz. These frequencies are in the radio frequency (RF) range, which are much lower than X-rays and do not disrupt living tissue.

If by some means the magnetic moment is displaced from the field, as shown in the Fig. 16.11, it will *precess* (rotate) around the field as a spinning top precesses in the gravitational field of the Earth. The frequency of precession is the Larmor frequency given by Eq. 16.4. The displacement of the magnetic moment is due to a reversal of alignment for some of the individual

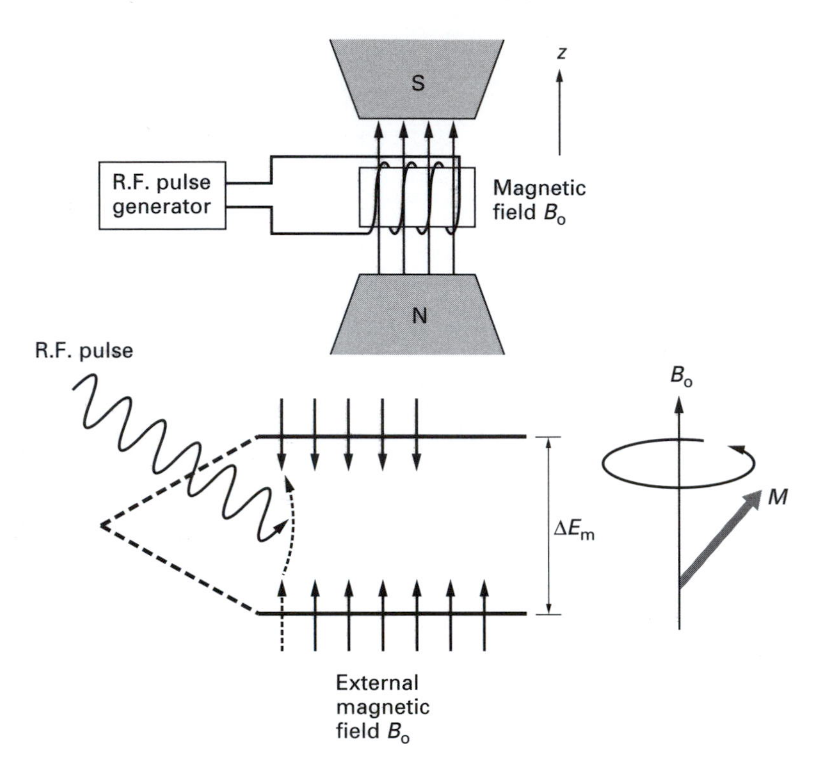

FIGURE 16.11 ▶ A short radio frequency driving pulse at the Larmor frequency displaces the magnetic moment from the external magnetic field by an angle determined by the strength and duration of the driving pulse.

nuclear magnetic moments from parallel to antiparallel alignment as shown in Fig. 16.11. A displacement of 90° corresponds to equalizing the population of the spin up and spin down states. To reverse the alignment of antiparallel spins requires energy which must be supplied by an external source.

The energy required to displace the magnetic moment from the direction of the external field is supplied by a short radio frequency driving pulse at the Larmor frequency which is the natural (resonant) frequency of precession. (This is analogous to setting a pendulum swinging by applying to it a force at the frequency of the pendulum resonance.) The driving pulse is applied by a coil surrounding the sample as shown in Fig. 16.11. At the end of the spin-flipping driving pulse, the magnetic moment is displaced from the external magnetic field by an angle determined by the strength and duration of the driving pulse.

The displaced magnetic moment produced by the radio frequency driving pulse, precesses around the external magnetic field and itself generates a radio frequency signal at the Larmor frequency of rotation. This emitted NMR signal can be detected by a separate coil or by the driving coil itself. The detected NMR signal decreases exponentially with time due to two distinct processes, (1) the return of the nuclear spin orientations to the equilibrium distribution and (2) variations in the local magnetic field.

Process 1: As was stated earlier, in the presence of an external field more of the nuclei are lined up parallel to the field than antiparallel. The radio frequency pulse flips some of the antiparallel spins into the parallel configuration. As soon as the driving pulse is over, the nuclear spins and the associated magnetic moment begin to return back to the original equilibrium alignment. The equilibration is brought about by the exchange of energy between the nuclear spins and the surrounding molecules. With the return of the magnetic moment to the original alignment with the external magnetic field, the precession angle decreases, as does the associated NMR signal. The decay of the NMR signal is exponential with time constant T_1, called the spin lattice relaxation time.

Process 2: The local magnetic field throughout the object under examination is not perfectly uniform. Variations in the magnetic field are produced by the magnetic properties of molecules adjacent to the nuclear spins. Such variations in the local magnetic field cause the Larmor frequency of the individual nuclear magnetic moments to differ slightly from each other. As a result, the precessions of the nuclei get out of phase with each other, and the total NMR signal decreases. This dephasing is likewise exponential, with a time constant T_2, called the spin-spin relaxation time.

The driving pulse and the emitted NMR signal are shown schematically in Fig. 16.12. The NMR signal detected after the driving pulse contains information about the material being studied. For a given initial driving pulse, the magnitude of the emitted NMR signal is a function of the number of hydrogen nuclei in the material. Bone, for example, which contains relatively few water or other hydrogen-containing molecules, produces a relatively low NMR signal. The post-pulse radiation emitted by fatty tissue is much higher.

The time constants T_1 and T_2 characterizing the rate of decay of the emitted NMR signal provide information about the nature of the material within which the precessing nuclei are located. The spinning top provides a useful analogy. A well-designed top in vacuum will spin for a long time. In air, the duration of the spin will be somewhat shorter because collisions with air molecules will dissipate its rotational energy. In water, where the frictional losses are yet greater, the top will spin hardly at all. The decay rate of the spinning top provides information about the nature of the medium surrounding the top. Similarly, the characteristic time constants T_1 and T_2 provide

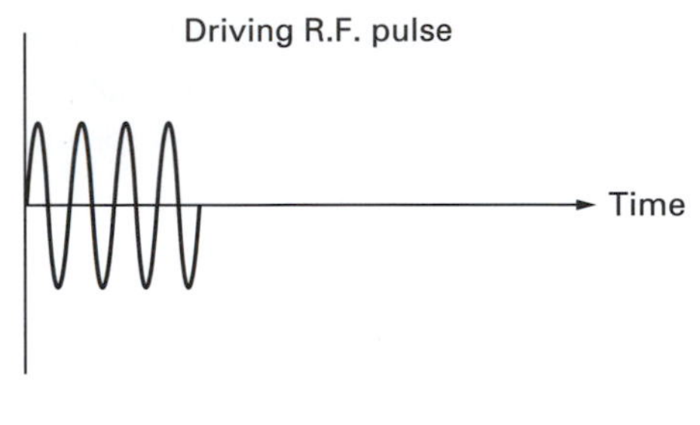

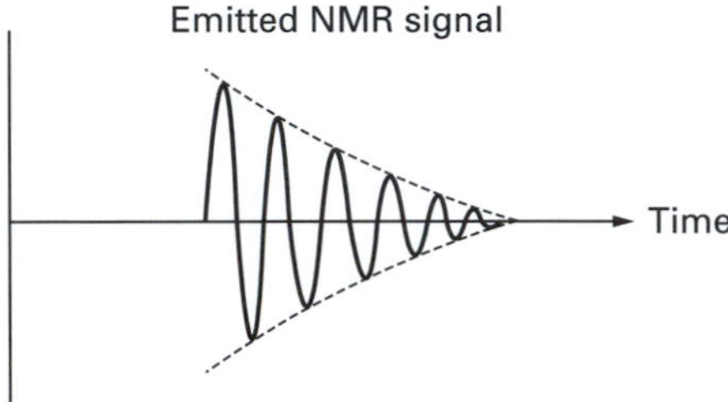

FIGURE 16.12 ▶ The driving pulse and the emitted NMR signal.

information about the matter surrounding the precessing nuclei. (See [16.4].) For example, with an external magnetic field of 1 T, for fat $T_1 = 240$ msec and $T_2 = 80$ msec; for heart tissue $T_1 = 570$ msec and $T_2 = 57$ msec. (See [16.4].) Malignant tissue is often characterized by higher values of T_1.

The NMR principles described have been used since the 1940s to identify molecules in various physics, chemistry, and biological applications. In this application the detected NMR signal is derived from the entire volume exposed to the magnetic field. The technique as discussed so far cannot provide information about the location of the signal within the volume studied.

16.11.2 Imaging with NMR

In order to obtain a three-dimensional image using nuclear magnetic resonance, we must isolate and identify the location of signals from small sections of the body and then build the image from these individual signals. In CT scans, such tomographic spatial images are obtained by extracting the information from intersection points of narrowly focused X-ray beams. This cannot be done with NMR because the wavelengths of the radio frequency driving signals are long, in the range of meters, which cannot be collimated into the narrow beams required to examine small regions of interest.

In the 1970s, several new techniques were developed to utilize NMR signals for the construction of two-dimensional tomographic images similar to those provided by CT scans. One of the first of these was described by P. C.

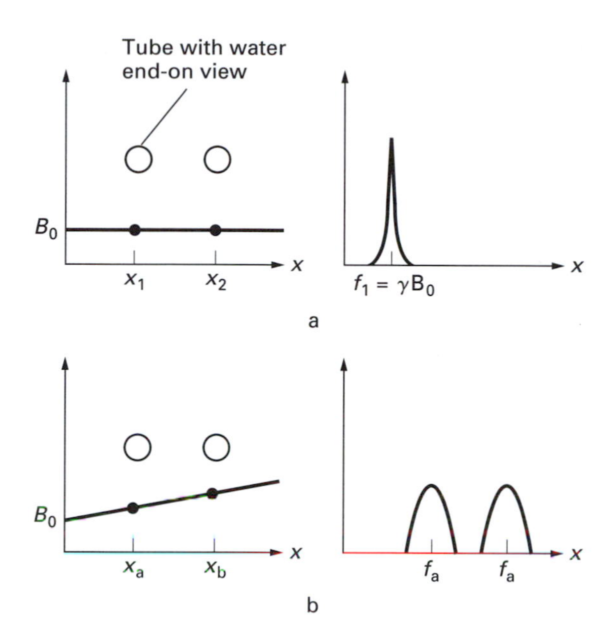

FIGURE 16.13 ▶ (a) In a uniform magnetic field (B_0) the Larmor frequency of two locations in space A and B is the same. (b) When a magnetic field gradient is superimposed on the uniform field, the Larmor frequencies at the locations A and B are different.

Lauterbur in 1973. He demonstrated the principle using two tubes of water, A and B as shown in Fig 16.13. In a uniform magnetic field (B_0) the Larmor frequency of the two tubes is the same. Therefore, the post-pulse NMR signals from tubes A and B cannot be distinguished. The NMR signals from the two tubes can be made distinguishable by superimposing on the uniform field B_0 a magnetic field gradient $B(x)$ as shown in Fig. 16.13b. The total magnetic field now changes with position along the x-axis, and the associated Larmor frequencies at the location of tubes A and B are now different.

As is evident, each point (actually small region Δx) on the x-axis is now characterized by its unique Larmor frequency. Therefore, the NMR signal observed after excitation with a pulse of a given frequency can be uniquely associated with a specific region in the x-space. A field gradient in one direction yields projection of the object onto that axis. To obtain a tomographic image in the x-y plane, a field gradient in both the x- and y-directions must be introduced. A magnetic field gradient is also applied in the z-direction to select within the body the slice to be examined. A very large number of NMR signals have to be collected and synthesized to construct an MRI image. For this purpose, the intensity as well as the time constants T_1 and T_2 of

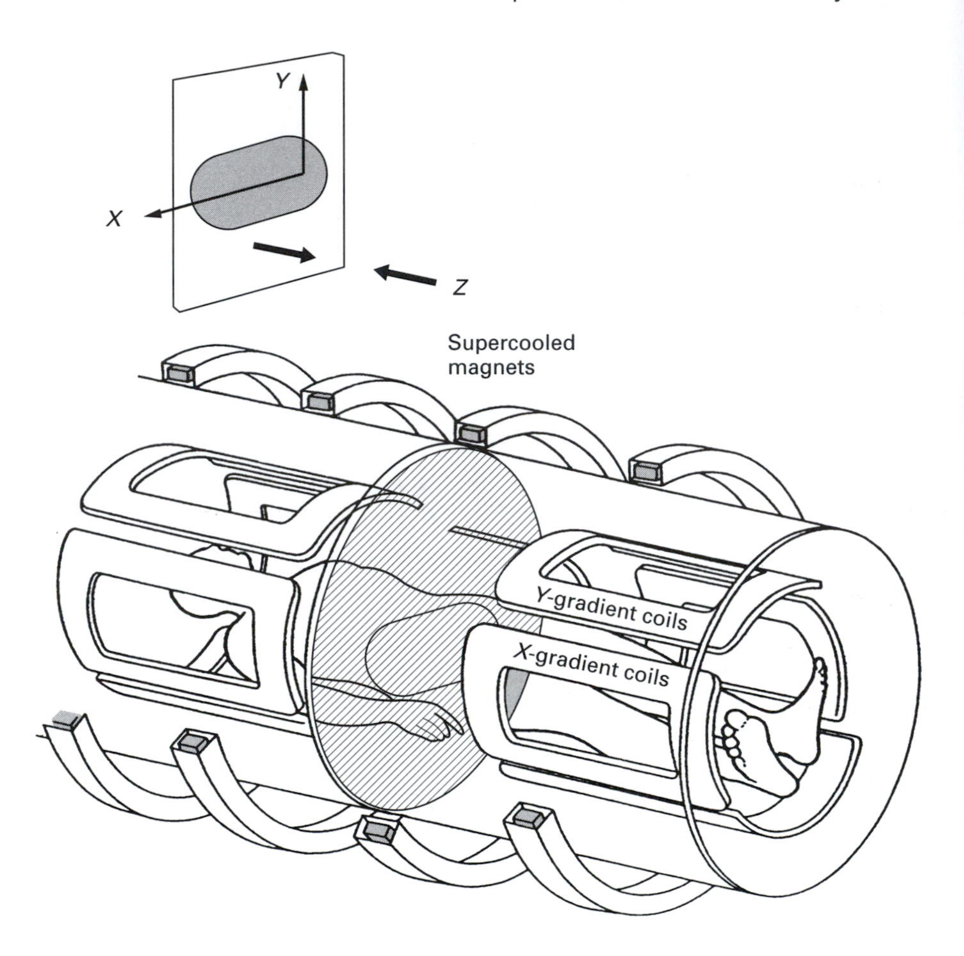

FIGURE 16.14 ▶ Sketch of a whole body MRI imaging system.

the NMR signal are needed. The process is more complex than for a CT scan and requires highly sophisticated computer programs.

A sketch of a whole-body MRI apparatus is shown in Fig. 16.14. Most such devices use liquid helium cooled superconducting magnets to produce the high magnetic fields required for the production of high resolution images. An MRI image of the brain is shown in Fig. 16.15.

16.12 Atomic Theory and Life

We have discussed in this book many phenomena in the life sciences that are clearly explained by the theories of physics. Now we come to the most

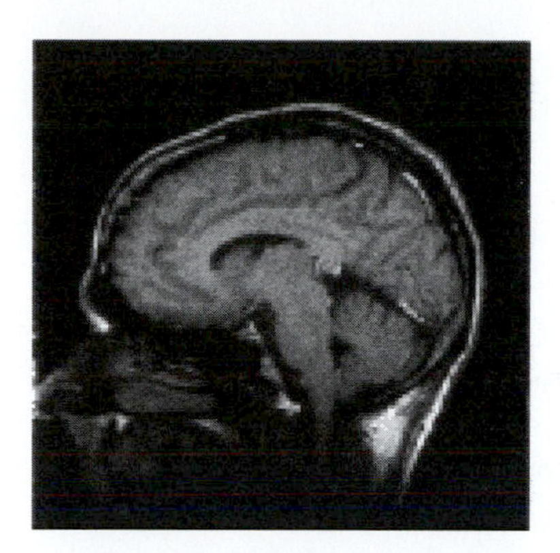

FIGURE 16.15 ▶ MRI image of brain. From V. Kuperman, "Magnetic Resonance Imaging," 2000, Academic Press.

fundamental question: Can physics explain life itself? In other words, if we put together the necessary combination of atoms, at each step following the known laws of physics, do we inevitably end up with a living organism, or must we invoke some new principles outside the realm of current physics in order to explain the occurrence of life? This is a very old question which still cannot be answered with certainty. But it can be clarified.

Quantum mechanics, which is the fundamental theory of modern atomic physics, has been very successful in describing the properties of atoms and the interaction of atoms with each other. Starting with a single proton and one electron, the theory shows that their interaction leads to the hydrogen atom with its unique configuration and properties. The quantum mechanical calculations for larger atoms are more complicated. In fact, so far a complete calculation has been performed only for the hydrogen atom. The properties of heavier atoms must be computed using various approximation techniques. Yet there is little doubt that quantum mechanics describes all the properties of atoms from the lightest to the heaviest. The experimental evidence gathered over the past 100 years fully confirms this view.

The interactions between atoms, which result in the formation of molecules, are likewise in the domain of quantum mechanics. Here again exact solutions of the quantum mechanical equations have been obtained only for the simplest molecule, H_2. Still it is evident that all the rules for both organic and inorganic chemistry follow from the principles of quantum mechanics. Even though our present numerical techniques cannot cope with the enormous calculations required to predict the exact configuration of a complex molecule, the concepts developed in physics and chemistry are applicable. The strengths of the interatomic bonds and the orientations of the atoms within the molecules are all in accord with the theory. This is true even for the largest organic molecules such as the proteins and DNA.

Past this point, however, we encounter a new level of organization: the cell. The organic molecules, which are in themselves highly complex, combine to form cells, which in turn are combined to form larger living organisms, which possess all the amazing properties of life. These organisms take nourishment from the environment, grow, reproduce, and at some level begin to govern their own actions. Here it is no longer obvious that the theories governing the interaction of atoms lead directly to these functions that characterize life. We are now in the realm of speculations.

The phenomena associated with life show such remarkable organization and planning that we may be tempted to suggest that perhaps some new undiscovered law governs the behavior of organic molecules that come together to form life. Yet there is no evidence for any special laws operating within living systems. So far, on all levels of examination, the observed phenomena associated with life obey the well-known laws of physics. This does not mean that the existence of life follows from the basic principles of physics, but it may. In fact the large organic molecules inside cells are sufficiently complex to contain within their structures the information necessary to guide in a predetermined way the activities associated with life. Some of these codes contained in the specific groupings of atoms within the molecules have now been unraveled. Because of these specific structures, a given molecule always participates in a well-defined activity within the cell. It is very likely that all the complex functions of cells and of cell aggregates are simply the collective result of the enormously large number of predetermined but basically well-understood chemical reactions.

This still leaves the most important question unanswered: What are the forces and the principles that initially cause the atoms to assemble into coded molecules which then ultimately lead to life. The answer here is probably again within the scope of our existing theories of matter.

In 1951, S. L. Miller simulated in his laboratory the type of conditions that may have existed perhaps 3.5 billion years ago in the atmosphere of the primordial Earth. He circulated a mixture of water, methane, ammonia, and hydrogen through an electric discharge. The discharge simulated the energy sources that were then available from the sun, lightning, and radioactivity. After about one week Miller found that the chemical activities in the mixture produced organic molecules including some of the simple amino acids, which are the building blocks of proteins. Since then, hundreds of other organic molecules have been synthesized under similar conditions. Many of them resemble the components of the important large molecules found in cells. It is thus plausible that in the primordial oceans, rich in organic molecules produced by the prevailing chemical reactions, life began. A number of smaller organic molecules combined accidentally to form a large self-replicating molecule such as DNA. These, in turn, combined into organized aggregates and finally into living cells.

Although the probability for the spontaneous occurrence of such events is small, the time span of evolution is probably long enough to make this scenario plausible. If that is indeed the case, the current laws of physics can explain all of life. At the present state of knowledge about life processes, the completeness of the descriptions provided by physics cannot be proved. The principles of physics have certainly explained many phenomena, but mysteries remain. At present, however, there seems to be no need to invoke any new laws.

► EXERCISES ►

16-1. Explain the operation of a spectrometer and describe two possible uses for this device.

16-2. Describe the process of X-ray computerized tomography. What information does this process provide that ordinary X-ray images do not?

16-3. What is your (considered) opinion of food preservation by radiation?

16-4. Describe the basic principles of magnetic resonance imaging.

16-5. Discuss some of the most notable attributes of living systems that distinguish them from inanimate ones.

Appendix A

Basic Concepts in Mechanics

In this section, we will define some of the fundamental concepts in mechanics. We assume that the reader is familiar with these concepts and that here a simple summary will be sufficient. A detailed discussion can be found in basic physics texts, some of which are listed in the Bibliography (Appendix D).

A.1 Speed and Velocity

Velocity is defined as the rate of change of position with respect to time. Both magnitude and direction are necessary to specify velocity. Velocity is, therefore, a vector quantity. The magnitude of the velocity is called *speed*. In the special case when the velocity of an object is constant, the distance s traversed in time t is given by

$$s = vt \tag{A.1}$$

In this case, velocity can be expressed as

$$v = \frac{s}{t} \tag{A.2}$$

If the velocity changes along the path, the expression s/t yields the average velocity.

A.2 Acceleration

If the velocity of an object along its path changes from point to point, its motion is said to be *accelerated* (or decelerated). Acceleration is defined as the rate of change in velocity with respect to time. In the special case of uniform acceleration, the final velocity v of an object that has been accelerated for a time t is

$$v = v_0 + at \tag{A.3}$$

Here v_0 is the initial velocity of the object, and a is the acceleration.[1] Acceleration can, therefore, be expressed as

$$a = \frac{v - v_0}{t} \tag{A.4}$$

In the case of uniform acceleration, a number of useful relations can be simply derived. The average velocity during the interval t is

$$v_{\mathrm{av}} = \frac{v + v_0}{2} \tag{A.5}$$

The distance traversed during this time is

$$s = v_{\mathrm{av}} t \tag{A.6}$$

Using Eqs. A.4 and A.5, we obtain

$$s = v_0 t + \frac{at^2}{2} \tag{A.7}$$

By substituting $t = (v - v_0)/a$ (from Eq. A.4) into Eq. A.7, we obtain

$$v^2 = v_0^2 + 2as \tag{A.8}$$

A.3 Force

Force is a push or a pull exerted on a body which tends to change the state of motion of the body.

A.4 Pressure

Pressure is the force applied to a unit area.

---

[1] Both velocity and acceleration may vary along the path. In general, velocity is defined as the time derivative of the distance along the path of the object; that is,

$$v = \lim_{\Delta t \to 0} \frac{\Delta s}{\Delta t} = \frac{ds}{dt}$$

Acceleration is defined as the time derivative of the velocity along the path; that is,

$$a = \frac{dv}{dt} = \frac{d}{dt}\left(\frac{ds}{dt}\right) = \frac{d^2 s}{dt^2}$$

A.5 Mass

We have stated that a force applied to a body tends to change its state of motion. All bodies have the property of resisting change in their motion. Mass is a quantitative measure of inertia or the resistance to a change in motion.

A.6 Weight

Every mass exerts an attractive force on every other mass; this attraction is called the *gravitational force*. The weight of a body is the force exerted on the body by the mass of the Earth. The weight of a body is directly proportional to its mass. Weight being a force is a vector, and it points vertically down in the direction of a suspended plumb line.

Mass and weight are related but distinct properties of an object. If a body were isolated from all other bodies, it would have no weight, but it would still have mass.

A.7 Linear Momentum

Linear momentum of a body is the product of its mass and velocity; that is,

$$\text{Linear momentum} = mv \tag{A.9}$$

A.8 Newton's Laws of Motion

The foundations of mechanics are Newton's three *laws of motion*. The laws are based on observation, and they cannot be derived from more basic principles. These laws can be stated as follows:

First Law: *A body remains at rest or in a state of uniform motion in a straight line unless it is acted on by an applied force.*

Second Law: *The time rate of change of the linear momentum of a body is equal to the force F applied to it.*
Except at very high velocities, where relativistic effects must be considered, the second law can be expressed mathematically in terms of the

mass m and acceleration a of the object as[2]

$$F = ma \tag{A.10}$$

This is one of the most commonly used equations in mechanics. It shows that if the applied force and the mass of the object are known, the acceleration can be calculated. When the acceleration is known, the velocity of the object and the distance traveled can be computed from the previously given equations.

The Earth's gravitational force, like all other forces, causes an acceleration. By observing the motion of freely falling bodies, this acceleration has been measured. Near the surface of the Earth, it is approximately 9.8 m/sec^2. Because gravitational acceleration is frequently used in computations, it has been given a special symbol g. Therefore, the gravitational force on an object with mass m is

$$F_{\text{gravity}} = mg \tag{A.11}$$

This is, of course, also the weight of the object.

Third Law: *For every action, there is an equal and opposite reaction.* This law implies that when two bodies A and B interact so that A exerts a force on B, a force of the same magnitude but opposite in direction is exerted by B on A. A number of illustrations of the third law are given in the text.

A.9 Conservation of Linear Momentum

It follows from Newton's laws that the total linear momentum of a system of objects remains unchanged unless acted on by an outside force.

A.10 Radian

In the analysis of rotational motion, it is convenient to measure angles in a unit called a *radian*. With reference to Fig. A.1, the angle in radian units is defined as

$$\theta = \frac{s}{r} \tag{A.12}$$

---

[2]The second law can be expressed mathematically in terms of the time derivative of momentum: that is,

$$\text{Force} = \Big|_{\Delta t \to 0} \frac{mv\,(t + \Delta t) - mv(t)}{\Delta t} = \frac{d}{dt}\,(mv) = m\frac{dv}{d:} = ma$$

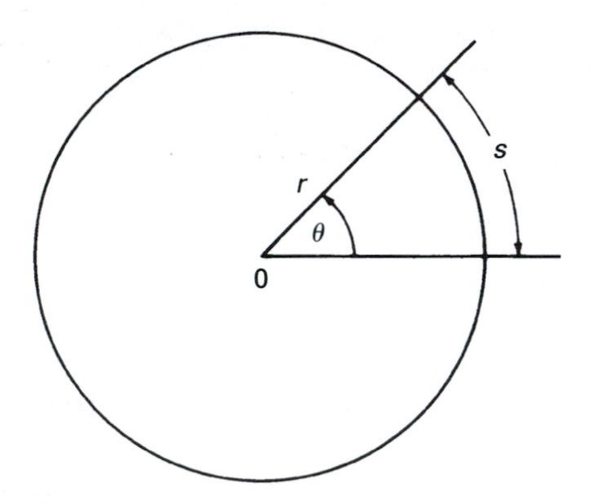

FIGURE A.1 ▶ The radian.

where s is the length of the circular arc and r is the radius of rotation. In a full circle, the arc length is the circumference $2\pi r$. Therefore in radian units the angle in a full circle is

$$\theta = \frac{2\pi r}{r} = 2\pi \text{ rad}$$

Hence,

$$1 \text{ rad} = \frac{360^\circ}{2\pi} = 57.3^\circ$$

A.11 Angular Velocity

The angular velocity ω is the angular displacement per unit time; that is, if a body rotates through an angle θ (in radians) in a time t, the angular velocity is

$$\omega = \frac{\theta}{t}(\text{rad/sec}) \qquad \text{(A.13)}$$

A.12 Angular Acceleration

Angular acceleration α is the time rate of change of angular velocity. If the initial angular velocity is ω_0 and the final angular velocity after a time t is ω_f, the angular acceleration is[3]

$$\alpha = \frac{\omega_f - \omega_0}{t} \tag{A.14}$$

A.13 Relations between Angular and Linear Motion

As an object rotates about an axis, each point in the object travels along the circumference of a circle; therefore, each point is also in linear motion. The linear distance s traversed in angular motion is

$$s = r\theta$$

The linear velocity v of a point that is rotating at an angular velocity ω a distance r from the center of rotation is

$$v = r\omega \tag{A.15}$$

The direction of the vector v is at all points tangential to the path s. The linear acceleration along the path s is

$$a = r\alpha \tag{A.16}$$

A.14 Equations for Angular Momentum

The equations for angular motion are analogous to the equations for translational motion. For a body moving with a constant angular acceleration α and initial angular velocity ω_0, the relationships are shown in Table A.1.

A.15 Centripetal Acceleration

As an object rotates uniformly around an axis, the magnitude of the linear velocity remains constant, but the direction of the linear velocity is continuously changing. The change in velocity always points toward the center of

---

[3]Both angular velocity and angular acceleration may vary along the path. In general, the instantaneous angular velocity and acceleration are defined as

$$\omega = \frac{d\theta}{dt}; \qquad \alpha = \frac{d\omega}{dt} = \frac{d^2\theta}{dt^2}$$

TABLE A.1 ▶ Equations for Rotational Motion (angular acceleration, α = constant)

$$\omega = \omega_0 + \alpha t$$
$$\theta = \omega_0 t + \tfrac{1}{2}\alpha t^2$$
$$\omega^2 = \omega_0^2 + 2\alpha\theta$$
$$\omega_{av} = \frac{(\omega_0 + \omega)}{2}$$

rotation. Therefore, a rotating body is accelerated toward the center of rotation. This acceleration is called *centripetal* (center-seeking) *acceleration*. The magnitude of the centripetal acceleration is given by

$$a_c = \frac{v^2}{r} = \omega^2 r \tag{A.17}$$

where r is the radius of rotation and v is the speed tangential to the path of rotation. Because the body is accelerated toward its center of rotation, we conclude from Newton's second law that a force pointing toward the center of rotation must act on the body. This force, called the *centripetal force* F_c, is given by

$$F_c = ma_c = \frac{mv^2}{r} = m\omega^2 r \tag{A.18}$$

where m is the mass of the rotating body.

For a body to move along a curved path, a centripetal force must be applied to it. In the absence of such a force, the body moves in a straight line, as required by Newton's first law. Consider, for example, an object twirled at the end of a rope. The centripetal force is applied by the rope on the object. From Newton's third law, an equal but opposite reaction force is applied on the rope by the object. The reaction to the centripetal force is called the *centrifugal force*. This force is in the direction away from the center of rotation. The centripetal force, which is required to keep the body in rotation, always acts perpendicular to the direction of motion and, therefore, does no work. (See Eq. A.28.) In the absence of friction, energy is not required to keep a body rotating at a constant angular velocity.

A.16 Moment of Inertia

The moment of inertia in angular motion is analogous to mass in translational motion. The moment of inertia I of an element of mass m located a distance

TABLE A.2 ► Moments of Inertia of Some Simple Bodies

Body	Location of axis	Moment of inertia
A thin rod of length l	Through the center	$ml^2/12$
A thin rod of length l	Through one end	$ml^2/3$
Sphere of radius r	Along a diameter	$2mr^2/5$
Cylinder of radius r	Along axis of symmetry	$mr^2/2$

r from the center of rotation is

$$I = mr^2 \tag{A.19}$$

In general, when an object is in angular motion, the mass elements in the body are located at different distances from the center of rotation. The total moment of inertia is the sum of the moments of inertia of the mass elements in the body.

Unlike mass, which is a constant for a given body, the moment of inertia depends on the location of the center of rotation. In general, the moment of inertia is calculated by using integral calculus. The moments of inertia for a few objects useful for our calculations are shown in Table A.2.

A.17 Torque

Torque is defined as the tendency of a force to produce rotation about an axis. Torque, which is usually designated by the letter L, is given by the product of the perpendicular force and the distance d from the point of application to the axis of rotation; that is (see Fig. A.2),

$$L = F\cos\theta \times d \tag{A.20}$$

The distance d is called the *lever arm* or *moment arm.*

A.18 Newton's Laws of Angular Motion

The laws governing angular motion are analogous to the laws of translational motion. Torque is analogous to force, and the moment of inertia is analogous to mass.

First Law: A body in rotation will continue its rotation with a constant angular velocity unless acted upon by an external torque.

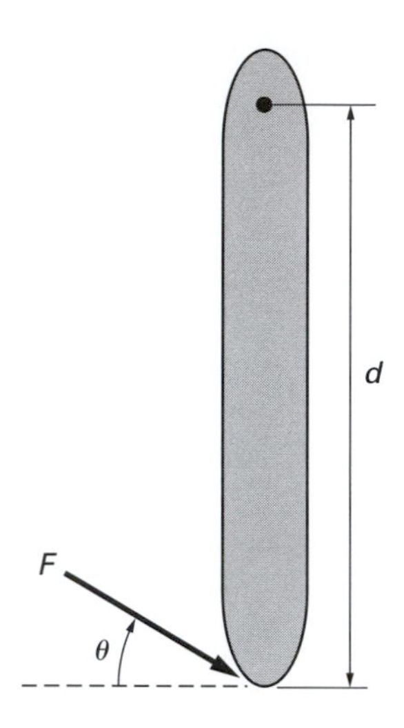

FIGURE A.2 ▶ Torque produced by a force.

Second Law: The mathematical expression of the second law in angular motion is analogous to Eq. A.10. It states that the torque is equal to the product of the moment of inertia and the angular acceleration; that is,

$$L = I\alpha \tag{A.21}$$

Third Law: For every torque, there is an equal and opposite reaction torque.

A.19 Angular Momentum

Angular momentum is defined as

$$\text{Angular momentum} = I\omega \tag{A.22}$$

From Newton's laws, it can be shown that angular momentum of a body is conserved if there is no unbalanced external torque acting on the body.

A.20 Addition of Forces and Torques

Any number of forces and torques can be applied simultaneously to a given object. Because forces and torques are vectors, characterized by both a magnitude and a direction, their net effect on a body is obtained by vectorial addition. When it is required to obtain the total force acting on a body, it is often convenient to break up each force into mutually perpendicular components.

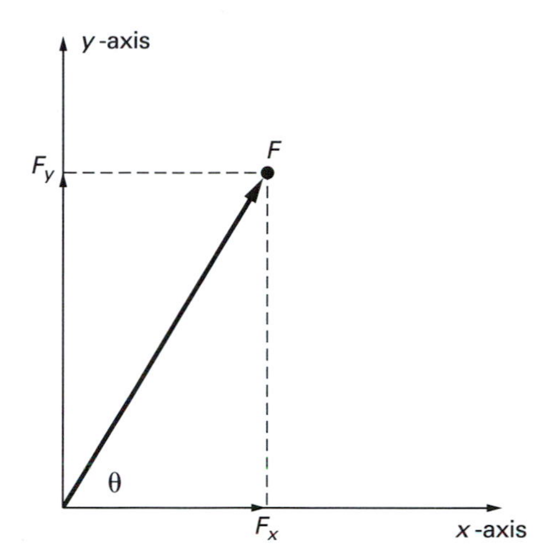

FIGURE A.3 ▶ The resolution of a force into its vertical and horizontal components.

This is illustrated for the two-dimensional case in Fig. A.3. Here we have chosen the horizontal x- and the vertical y-directions as the mutually perpendicular axes. In a more general three-dimensional case, a third axis is required for the analysis.

The two perpendicular components of the force F are

$$\begin{aligned} F_x &= F\cos\theta \\ F_y &= F\sin\theta \end{aligned} \tag{A.23}$$

The magnitude of the force F is given by

$$F = \sqrt{F_x^2 + F_y^2} \tag{A.24}$$

When adding a number of forces (F_1, F_2, F_3, ...) the mutually perpendicular components of the total force F_T are obtained by adding the corresponding components of each force; that is,

$$\begin{aligned} (F_T)_x &= (F_1)_x + (F_2)_x + (F_3)_x + \cdots \\ (F_T)_y &= (F_1)_y + (F_2)_y + (F_3)_y + \cdots \end{aligned} \tag{A.25}$$

The magnitude of the total force is

$$F_T = \sqrt{(F_T)_x^2 + (F_T)_y^2} \tag{A.26}$$

The torque produced by a force acts to produce a rotation in either a clockwise or a counterclockwise direction. If we designate one direction of rotation as positive and the other as negative, the total torque acting on a body is obtained by the addition of the individual torques each with the appropriate sign.

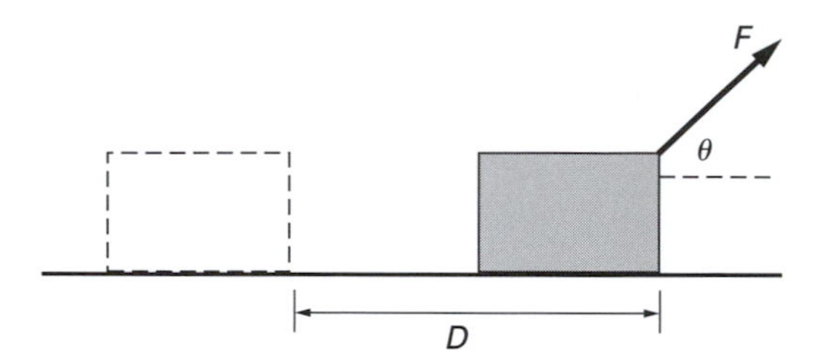

FIGURE A.4 ▶ Work done by a force.

A.21 Static Equilibrium

A body is in static equilibrium if both its linear and angular acceleration are zero. To satisfy this condition, the sum of the forces F acting on the body, as well as the sum of the torques L produced by these forces must be zero; that is,

$$\sum F = 0 \qquad \text{and} \qquad \sum L = 0 \tag{A.27}$$

A.22 Work

In our everyday language, the word *work* denotes any types of effort whether physical or mental. In physics, a more rigorous definition is required. Here work is defined as the product of force and the distance through which the force acts. Only the force parallel to the direction of motion does work on the object. This is illustrated in Fig. A.4. A force F applied at an angle θ pulls the object along the surface through a distance D. The work done by the force is

$$\text{Work} = F\cos\theta \times D \tag{A.28}$$

A.23 Energy

Energy is an important concept. We find reference to energy in connection with widely different phenomena. We speak of atomic energy, heat energy, potential energy, solar energy, chemical energy, kinetic energy; we even speak of people as being full of energy. The common factor that ties together these manifestations is the possibility of obtaining work from these sources. The connection between energy and work is simple: Energy is required to do work. Energy is measured in the same units as work; in fact, there is a one-to-one correspondence between them. It takes 2 J of energy to do 2 J of work. In all physical processes, energy is conserved. Through work, one form of energy can be converted into another, but the total amount of energy remains unchanged.

A.24 Forms of Energy

A.24.1 Kinetic Energy

Objects in motion can do work by virtue of their motion. For example, when a moving object hits a stationary object, the stationary object is accelerated. This implies that the moving object applied a force on the stationary object and performed work on it. The kinetic energy (KE) of a body with mass m moving with a velocity v is

$$KE = \frac{1}{2}mv^2 \tag{A.29}$$

In rotational motion, the kinetic energy is

$$KE = \frac{1}{2}I\omega^2 \tag{A.30}$$

A.24.2 Potential Energy

Potential energy of a body is the ability of the body to do work because of its position or configuration. A body of weight W raised to a height H with respect to a surface has a potential energy (PE)

$$PE = WH \tag{A.31}$$

This is the amount of work that had to be performed to raise the body to height H. The same amount of energy can be retrieved by lowering the body back to the surface.

A stretched or compressed spring possesses potential energy. The force required to stretch or compress a spring is directly proportional to the length of the stretch or compression (s); that is,

$$F = ks \tag{A.32}$$

Here k is the spring constant. The potential energy stored in the stretched or compressed spring is

$$PE = \frac{1}{2}ks^2 \tag{A.33}$$

A.24.3 Heat

Heat is a form of energy, and as such it can be converted to work and other forms of energy. Heat, however, is not equal in rank with other forms of energy. While work and other forms of energy can be completely converted to heat, heat energy can only be converted partially to other forms of energy. This property of heat has far-reaching consequences which are discussed in Chapter 10.

Heat is measured in calorie units. One calorie (cal) is the amount of heat required to raise the temperature of 1 g of water by 1 C°. The heat energy required to raise the temperature of a unit mass of a substance by 1 degree is called the *specific heat*. One calorie is equal to 4.184 J.

A heat unit frequently used in chemistry and in food technology is the *kilocalorie* or Cal which is equal to 1000 cal.

A.25 Power

The amount of work done—or energy expended—per unit time is called *power*. The algebraic expression for power is

$$P = \frac{\Delta E}{\Delta t} \tag{A.34}$$

where ΔE is the energy expended in a time interval Δt.

A.26 Units and Conversions

In our calculations we will mostly use SI units in which the basic units for length, mass, and time are meter, kilogram, and second. However, other units are also encountered in the text. Units and conversion factors for the most commonly encountered quantities are listed here with their abbreviations.

A.26.1 Length

SI unit: meter (m)

Conversions: 1 m = 100 cm (centimeter) = 1000 mm (millimeter)
1000 m = 1 km
1 m = 3.28 feet = 39.37 in
1 km = 0.621 mile
1 in = 2.54 cm

In addition, the micron and the angstrom are used frequently in physics and biolog

1 micron (μm) = 10^{-6} m = 10^{-4} cm
1 angstrom (Å)* = 10^{-8} cm

A.26.2 Mass

SI unit: kilogram (kg)
Conversions: 1 kg = 1000 g
The weight of a 1-kg mass is 9.8 newton (N).

A.26.3 Force

SI Unit: kg m s^{-2}, name of unit: newton (N)
Conversions: 1 N = 10^5 dynes (dyn) = 0.225 lbs

A.26.4 Pressure

SI unit: kg m^{-1} s^{-2}, name of unit: pascal (Pa)
Conversions: 1 Pa = 10^{-1} dynes/cm^2 = 9.87×10^{-6} atmosphere (atm)
= 1.45×10^{-4} lb/in^2
1 atm = 1.01×10^5 Pa = 760 mmHg (torr)

A.26.5 Energy

SI unit: kg m^{-2} s^{-2}, name of unit: joule (J)
Conversion: 1 J = 1 N-m = 10^7 ergs = 0.239 cal = 0.738 ft-lb

A.26.6 Power

SI unit: J s^{-1}, name of unit: watt (W)
Conversion: 1 W = 10^7 ergs/sec = 1.34×10^{-3} horsepower (hp)

Appendix B

Review of Electricity

B.1 Electric Charge

Matter is composed of atoms. An atom consists of a nucleus surrounded by electrons. The nucleus itself is composed of protons and neutrons. Electric charge is a property of protons and electrons. There are two types of electric charge: positive and negative. The proton is positively charged, and the electron is negatively charged. All electrical phenomena are due to these electric charges.

Charges exert forces on each other. Unlike charges attract and like charges repel each other. The electrons are held around the nucleus by the electrical attraction of the protons. Although the proton is about 2000 times heavier than the electron, the magnitude of the charge on the two is the same. There are as many positively charged protons in an atom as negatively charged electrons. The atom as a whole is, therefore, electrically neutral. The identity of an atom is determined by the number of protons in the nucleus. Thus, for example, hydrogen has 1 proton; nitrogen has 7 protons; and gold has 79 protons.

It is possible to remove electrons from an atom, making it positively charged. Such an atom with missing electrons is called a *positive ion*. It is also possible to add an electron to an atom which makes it a *negative ion*.

Electric charge is measured in coulombs (C). The magnitude of the charge on the proton and the electron is 1.60×10^{-19} C. The force F between two charged bodies is proportional to the product of their charges Q_1 and Q_2 and is inversely proportional to the square of the distance R between them; that is,

$$F = \frac{KQ_1Q_2}{R^2} \tag{B.1}$$

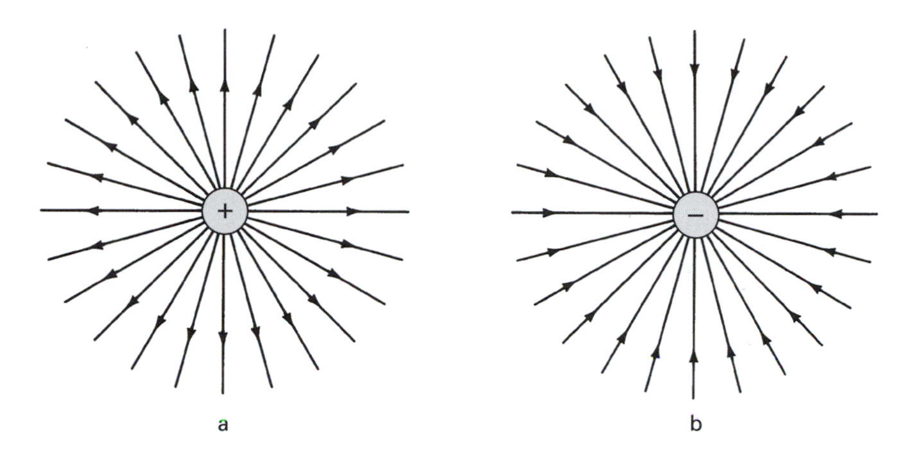

FIGURE B.1 ▶ Two-dimensional representation of the electric field produced by a positive point charge (a) and a negative point charge (b).

This equation is known as *Coulomb's law*. If R is measured in meters, the constant K is 9×10^9, and F is obtained in newtons.

B.2 Electric Field

An electric charge exerts a force on another electric charge; a mass exerts a force on another mass; and a magnet exerts a force on another magnet. All these forces have an important common characteristic: Exertion of the force does not require physical contact between the interacting bodies. The forces act at a distance. The concept of *lines of force* or *field lines* is useful in visualizing these forces which act at a distance.

Any object that exerts a force on another object without contact can be thought of as having lines of force emanating from it. The complete line configuration is called a *force field*. The lines point in the direction of the force, and their density at any point in space is proportional to the magnitude of the force at that point.

The lines of force emanate from an electric charge uniformly in all directions. By convention, the lines point in the direction of the force that the source charge exerts on a positive charge. Thus, the lines of force point away from a positive source charge and into a negative source charge (see Fig. B.1). The number of lines emanating from the charge is proportional to the magnitude of the electric charge. If the size of the source charge is doubled, the number of force lines is also doubled.

Lines of force need not be straight lines; as we mentioned, they point in the direction in which the force is exerted. As an example, we can consider the

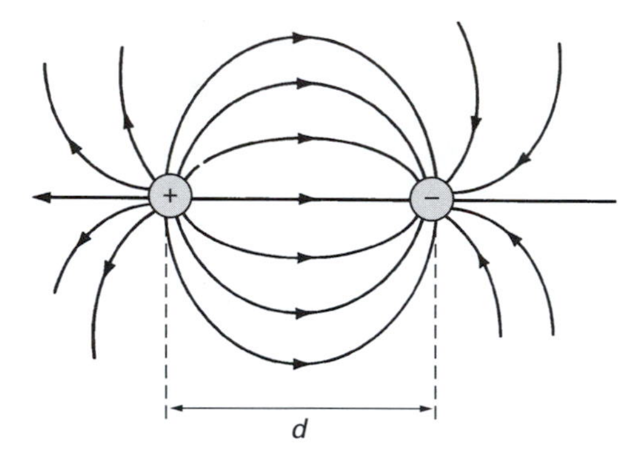

FIGURE B.2 ▶ Lines of force produced by a positive and a negative charge separated by a distance d.

net field due to two charges separated by a distance d. To determine this field we must compute the direction and size of the net force on a positive charge at all points in space. This is done by adding vectorially the force lines due to each charge. The force field due to a positive and negative charge of equal magnitude separated by a distance d from each other is shown in Fig. B.2. Here the lines of force are curved. This is, of course, the direction of the net force on a positive charge in the region surrounding the two fixed charges. The field shown in Fig. B.2 is called a *dipole field*, and it is similar to the field produced by a bar magnet.

B.3 Potential Difference or Voltage

The electric field is measured in units of volt per meter (or volt per centimeter). The product of the electric field and the distance over which the field extends is an important parameter which is called *potential difference* or *voltage*. The voltage (V) between two points is a measure of energy transfer as the charge moves between the two points. Potential difference is measured in volts. If there is a potential difference between two points, a force is exerted on a charge placed in the region between these points. If the charge is positive, the force tends to move it away from the positive point and toward the negative point.

B.4 Electric Current

An electric current is produced by a motion of charges. The magnitude of the current depends on the amount of charge flowing past a given point in a given

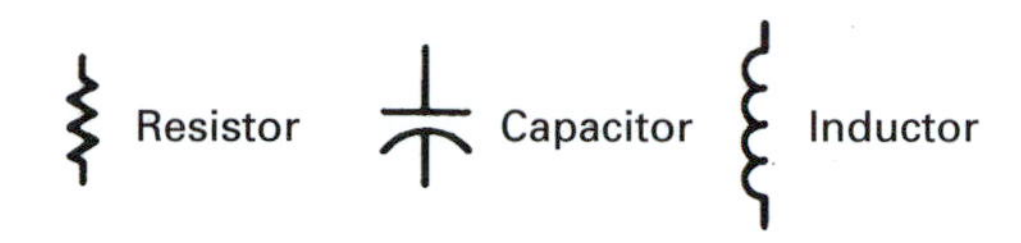

FIGURE B.3 ▶ Circuit components.

period of time. Current is measured in amperes (A). One ampere is 1 coulomb (C) of charge flowing past a point in 1 second (sec).

B.5 Electric Circuits

The amount of current flowing between two points in a material is proportional to the potential difference between the two points and to the electrical properties of the material. The electrical properties are usually represented by three parameters: resistance, capacitance, and inductance. Resistance measures the opposition to current flow. This parameter depends on the property of the material called *resistivity*, and it is analogous to friction in mechanical motion. Capacitance measures the ability of the material to store electric charges. Inductance measures the opposition in the material to changes in current flow. All materials exhibit to some extent all three of these properties; often, however, one of these properties is predominant. It is possible to manufacture components with specific values of resistance, capacitance, or inductance. These are called, respectively, *resistors*, *capacitors*, and *inductors*.

The schematic symbols for these three electrical components are shown in Fig. B.3. Electrical components can be connected together to form an electric circuit. Currents can be controlled by the appropriate choice of components and interconnections in the circuit. An example of an electric circuit is shown in Fig. B.4. Various techniques have been developed to analyze such circuits and to calculate voltages and currents at all the points in the circuit.

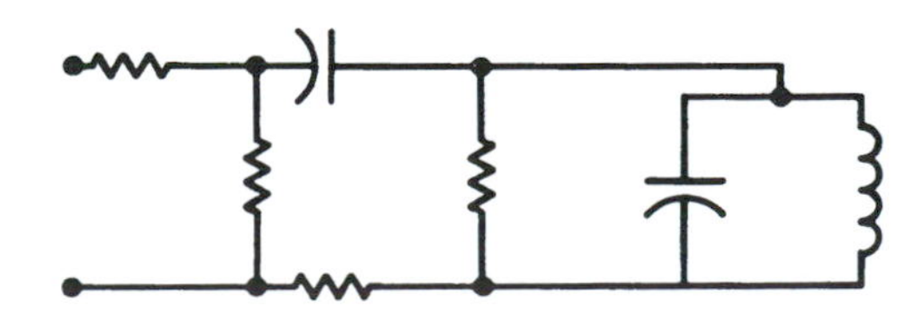

FIGURE B.4 ▶ Example of an electric circuit.

B.5.1 Resistor

The resistor is a circuit component that opposes current flow. Resistance (R) is measured in units of ohm (Ω). The relation between current (I) and voltage (V) is given by Ohm's law, which is

$$V = IR \tag{B.2}$$

Materials that present a very small resistance to current flow are called *conductors*. Materials with a very large resistance are called *insulators*. A flow of current through a resistor is always accompanied by power dissipation as electrical energy is converted to heat. The power (P) dissipated in a resistor is given by

$$P = I^2R \tag{B.3}$$

The inverse of resistance is called *conductance*, which is usually designated by the symbol G. Conductance is measured in units of *mho*, also called *siemens*. The relationship between conductance and resistance is

$$G = \frac{1}{R} \tag{B.4}$$

B.5.2 Capacitor

The capacitor is a circuit element that stores electric charges. In its simplest form it consists of two conducting plates separated by an insulator (see Fig. B.5). Capacitance (C) is measured in *farads*. The relation between the stored charge (Q), and the voltage across the capacitor is given by

$$Q = CV \tag{B.5}$$

In a charged capacitor, positive charges are on one side of the plate, and negative charges are on the other. The amount of energy (E) stored in such a

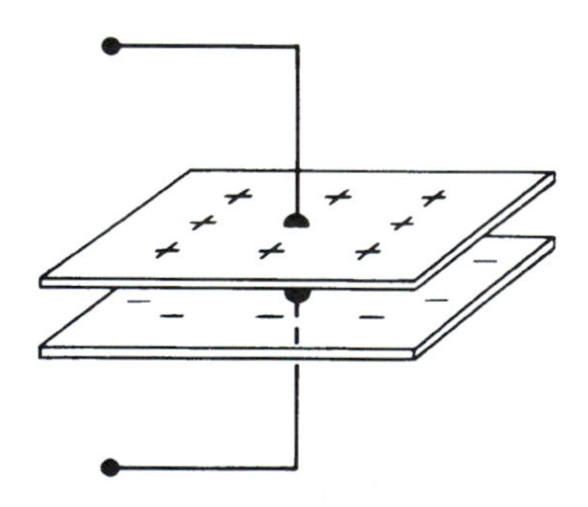

FIGURE B.5 ► A simple capacitor.

configuration is given by

$$E = \frac{1}{2}CV^2 \tag{B.6}$$

B.5.3 Inductor

The *inductor* is a device that opposes a change in the current flowing through it. Inductance is measured in units called *henry*.

B.6 Voltage and Current Sources

Voltages and currents can be produced by various batteries and generators. Batteries are based on chemical reactions that result in a separation of positive and negative charges within a material. Generators produce a voltage by the motion of conductors in magnetic fields. The circuit symbols for these sources are shown in Fig. B.6.

B.7 Electricity and Magnetism

Electricity and magnetism are related phenomena. A changing electric field always produces a magnetic field, and a changing magnetic field always produces an electric field. All electromagnetic phenomena can be traced to this basic interrelationship. A few of the consequences of this interaction follow:

1. An electric current always produces a magnetic field at a direction perpendicular to the current flow.
2. A current is induced in a conductor that moves perpendicular to a magnetic field.
3. An oscillating electric charge emits electromagnetic waves at the frequency of oscillation. This radiation propagates away from the source at the speed of light. Radio waves, light, and X-rays are examples of electromagnetic radiation.

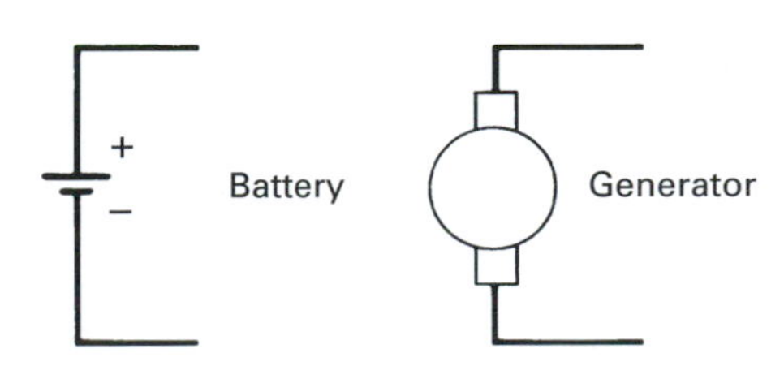

FIGURE B.6 ▶ Circuit symbols for a battery and a generator.

Appendix C

Review of Optics

C.1 Geometric Optics

The characteristics of optical components, such as mirrors and lenses, can be completely derived from the wave properties of light. Such detailed calculations, however, are usually rather complex because one has to keep track of the wave front along every point on the optical component. It is possible to simplify the problem if the optical components are much larger than the wavelength of light. The simplification entails neglecting some of the wave properties of light and considering light as a ray traveling perpendicular to the wave front (Fig. C.1). In a homogeneous medium, the ray of light travels in a straight line; it alters direction only at the interface between two media. This simplified approach is called *geometric optics*.

The speed of light depends on the medium in which it propagates. In vacuum, light travels at a speed of 3×10^8 m/sec. In a material medium, the speed of light is always less. The speed of light in a material is characterized by the index of refraction (n) defined as

$$n = \frac{c}{v} \tag{C.1}$$

where c is the speed of light in vacuum and v is the speed in the material. When light enters from one medium into another, its direction of propagation is changed (see Fig. C.2). This phenomenon is called *refraction*. The relationship between the angle of incidence (θ_1) and the angle of refraction (θ_2) is given by

$$\frac{\sin\theta_1}{\sin\theta_2} = \frac{n_2}{n_1} \tag{C.2}$$

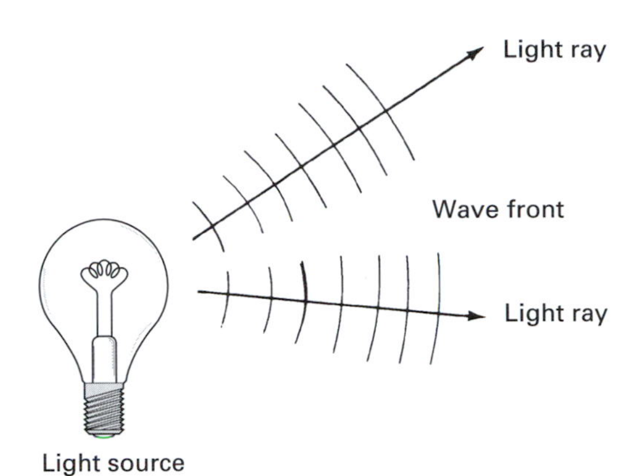

FIGURE C.1 ▶ Light rays perpendicular to the wave front.

The relationship in Eq. C.2 is called *Snell's law*. As shown in Fig. C.2, some of the light is also reflected. The angle of reflection is always equal to the angle of incidence.

In Fig. C.2a, the angle of incidence θ_1 for the entering light is shown to be greater than the angle of refraction θ_2. This implies that n_2 is greater than n_1 as would be the case for light entering from air into glass, for example. (See Eq. C.2.) If, on the other hand, the light originates in the medium of higher refractive index, as shown in Fig. C.2b, then the angle of incidence θ_1 is smaller than the angle of refraction θ_2. At a specific value of angle θ_1 called the *critical angle* (designated by the symbol θ_c), the light emerges tangent to the surface, that is, $\theta_2 = 90°$. At this point, $\sin\theta_2 = 1$ and, therefore, $\sin\theta_1 = \sin\theta_c = n_2/n_1$. Beyond this angle, that is for $\theta_1 > \theta_c$, light originating in the medium of higher refractive index does not emerge from the medium. At the interface, all the light is reflected back into the medium. This phenomenon is called *total internal reflection*. For glass, n_2 is typically 1.5, and the critical angle at the glass-air interface is $\sin\theta_c = 1/1.5$ or $\theta_c = 42°$.

Transparent materials such as glass can be shaped into lenses to alter the direction of light in a specific way. Lenses fall into two general categories: converging lenses and diverging lenses. A converging lens alters the direction of light so that the rays are brought together. A diverging lens has the opposite effect; it spreads the light rays apart.

Using geometric optics, we can calculate the size and shape of images formed by optical components, but we cannot predict the inevitable blurring of images which occurs as a result of the wave nature of light.

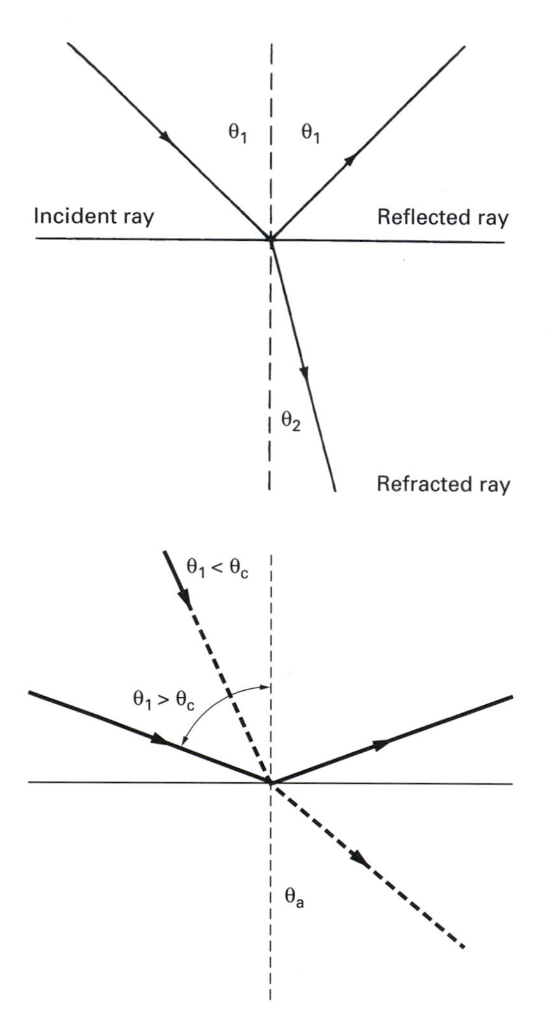

FIGURE C.2 ▶ (Top) Reflection and refraction of light. (Bottom) Total internal reflection.

C.2 Converging Lenses

A simple converging lens is shown in Fig. C.3. This type of a lens is called a convex lens.

Parallel rays of light passing through a convex lens converge at a point called the *principal focus of the lens*. The distance of this point from the lens is called the *focal length* f. Conversely, light from a point source at the focal point emerges from the lens as a parallel beam. The focal length of the lens

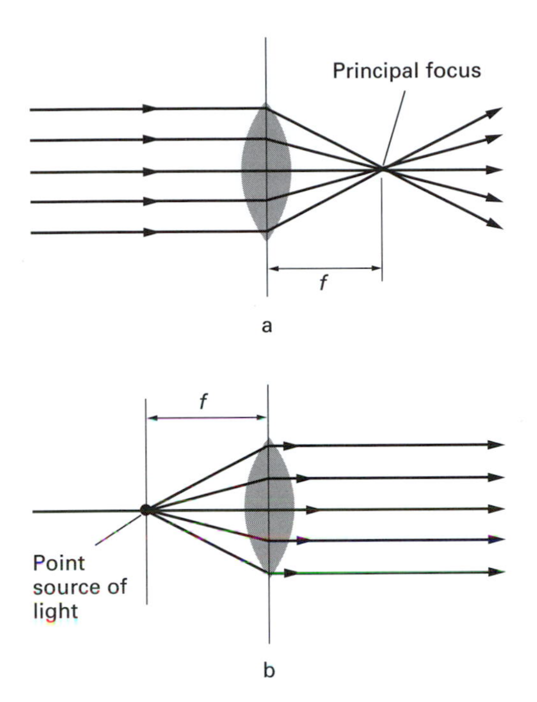

FIGURE C.3 ▶ The convex lens illuminated (a) by parallel light, (b) by point source at the focus.

is determined by the index of refraction of the lens material and the curvature of the lens surfaces. We adopt the following convention in discussing lenses.

1. Light travels from left to right.
2. The radius of curvature is positive if the curved surface encountered by the light ray is convex; it is negative if the surface is concave.

It can be shown that for a thin lens the focal length is given by

$$\frac{1}{f} = (n - 1)\left(\frac{1}{R_1} - \frac{1}{R_2}\right) \tag{C.3}$$

where R_1 and R_2 are the curvatures of the first and second surfaces, respectively (Fig. C.4). In Fig. C.4, R_2 is a negative number.

Focal length is a measure of the converging power of the lens. The shorter the focal length, the more powerful the lens. The focusing power of a lens is often expressed in diopters defined as

$$\text{Focusing power} = \frac{1}{f \text{ (meters)}} \text{ (diopters)} \tag{C.4}$$

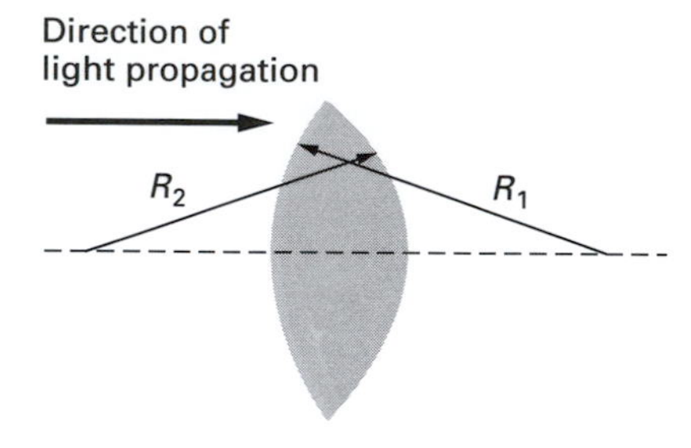

FIGURE C.4 ▶ Radius of curvature defined for a lens.

If two thin lenses with focal lengths f_1 and f_2, respectively, are placed close together, the focal length f_T of the combination is

$$\frac{1}{f_T} = \frac{1}{f_1} + \frac{1}{f_2} \tag{C.5}$$

Light from a point source located beyond the focal length of the lens is converged to a point image on the other side of the lens (Fig. C.5a). This type

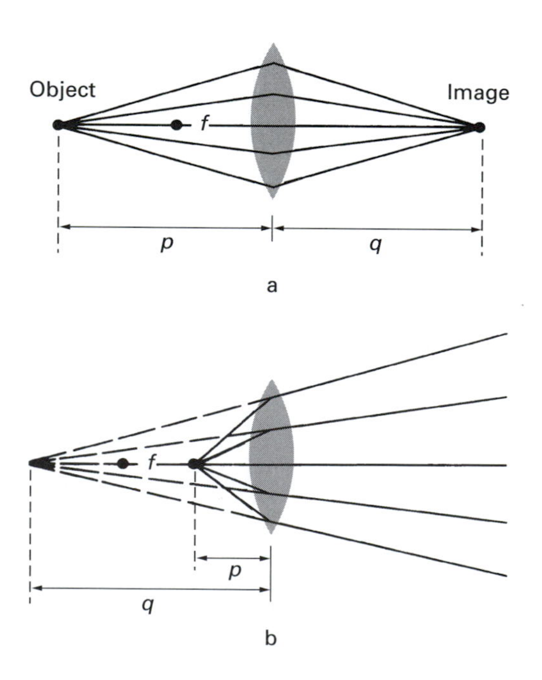

FIGURE C.5 ▶ Image formation by a convex lens: (a) real image, (b) virtual image.

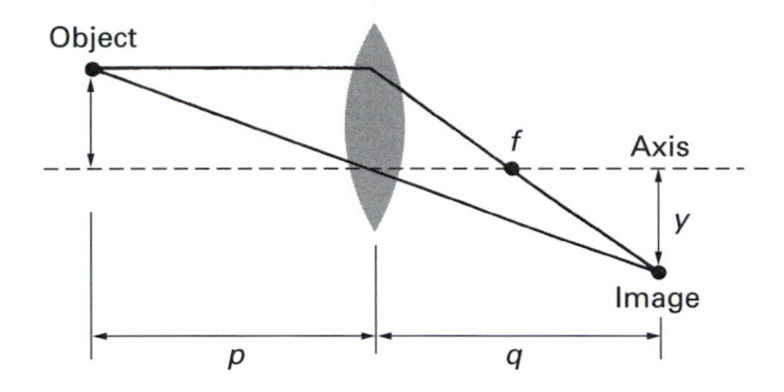

FIGURE C.6 ▶ Image formation off axis.

of an image is called a *real image* because it can be seen on a screen placed at the point of convergence.

If the distance between the source of light and the lens is less than the focal length, the rays do not converge. They appear to emanate from a point on the source side of the lens. This apparent point of convergence is called a *virtual image* (Fig. C.5b).

For a thin lens, the relationship between the source and the image distances from the lens is given by

$$\frac{1}{p} + \frac{1}{q} = \frac{1}{f} \tag{C.6}$$

Here p and q, respectively, are the source and the image distances from the lens. By convention, q in this equation is taken as positive if the image is formed on the side of the lens opposite to the source and negative if the image is formed on the source side.

Light rays from a source very far from the lens are nearly parallel; therefore, by definition we would expect them to be focused at the principal focal point of the lens. This is confirmed by Eq. C.6, which shows that as p becomes very large (approaches infinity), q is equal to f.

If the source is displaced a distance x from the axis, the image is formed at a distance y from the axis such that

$$\frac{y}{x} = \frac{q}{p} \tag{C.7}$$

This is illustrated for a real image in Fig. C.6. The relationship between p and q is still given by Eq. C.6.

C.3 Images of Extended Objects

So far we have discussed only the formation of images from point sources. The treatment, however, is easily applied to objects of finite size.

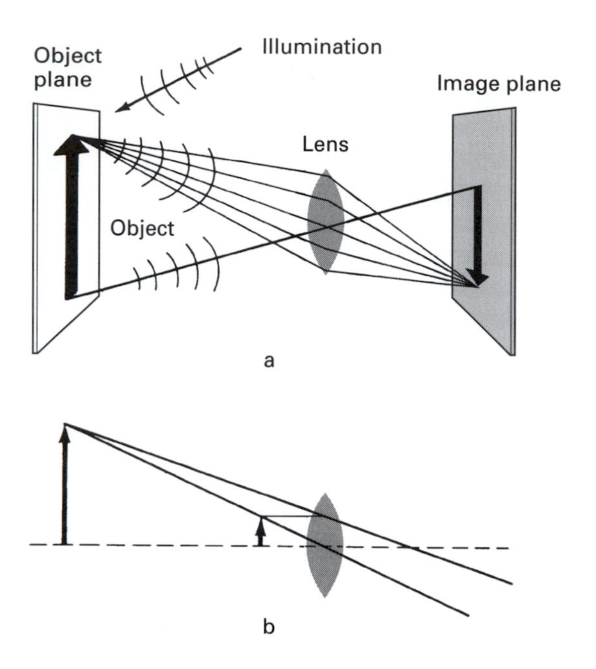

FIGURE C.7 ▶ Image of an object: (a) real, (b) virtual.

When an object is illuminated, light rays emanate from every point on the object (Fig. C.7a). Each point on the object plane a distance p from the lens is imaged at the corresponding point on the image plane a distance q from the lens. The relationship between the object and the image distances is given by Eq. C.6. As shown in Fig. C.7, real images are inverted and virtual images are upright. The ratio of image to object height is given by

$$\frac{\text{Image height}}{\text{Object height}} = -\frac{q}{p} \tag{C.8}$$

C.4 Diverging Lenses

An example of a diverging lens is the concave lens shown in Fig. C.8. Parallel light diverges after passing through a concave lens. The apparent source of origin for the diverging rays is the focal point of the concave lens. All the equations we have presented for the converging lens apply in this case also, provided the sign conventions are obeyed. From Eq. C.3, it follows that the

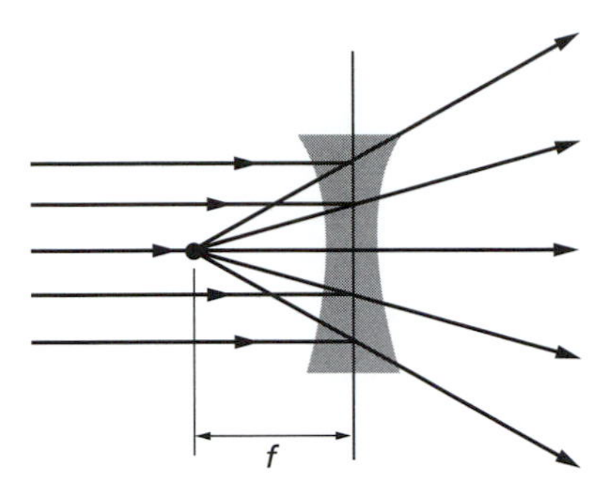

FIGURE C.8 ▶ A diverging lens.

focal length for a diverging lens is always negative and the lens produces only virtual images (Fig. C.8).

C.5 Lens Immersed in a Material Medium

The lens equations that we have presented so far apply in the case when the lens is surrounded by air that has a refraction index of approximately 1. Let us now consider the more general situation shown in Fig. C.9, which we will need in our discussion of the eye. The lens here is embedded in a medium that has a different index of refraction (n_1 and n_2) on each side of the lens. It can be shown (see [15-3]) that under these conditions the relationship between the object and the image distances is

$$\frac{n_1}{p} + \frac{n_2}{q} = \frac{n_L - n_1}{R_1} - \frac{n_L - n_2}{R_2} \tag{C.9}$$

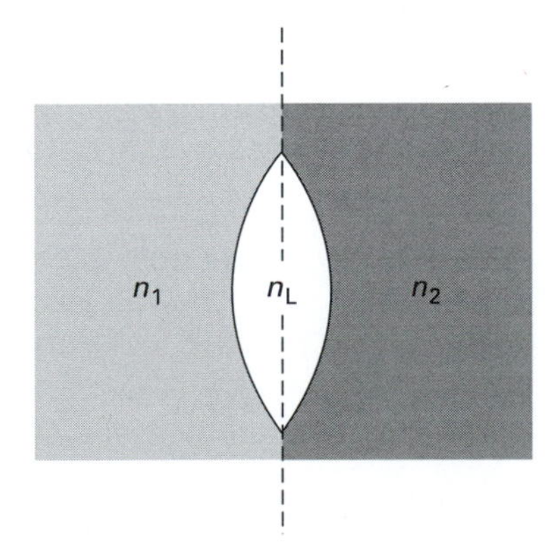

FIGURE C.9 ▶ Lens immersed in a material medium.

Here n_L is the refraction index of the lens material. The effective focal length in this case is

$$\frac{1}{f} = \frac{n_2 - n_1}{R_1} - \frac{n_L - n_2}{R_2} \tag{C.10}$$

Note that in air $n_1 = n_2 = 1$ and Eq. C.10 reduces to Eq. C.3.

The lens equations we have presented in this appendix assume that the lenses are thin. This is not a fully valid assumption for the lenses in the eye. Nevertheless these equations are adequate for our purposes.

Appendix D

Bibliography

D.1 Chapters 1 to 6

6-1 Alexander, R. McNeill. *Animal Mechanics*. London: Sidgwick and Jackson, 1968.

6-2 Baez, Albert V. *The New College Physics: A Spiral Approach*. San Francisco, CA: W. H. Freeman and Co., 1967.

6-3 Blesser, William B. *A Systems Approach to Biomedicine*. New York, NY: McGraw-Hill Book Co., 1969.

6-4 Bootzin, David, and Muffley, Harry C. *Biomechanics*. New York, NY: Plenum Press, 1969.

6-5 Cameron, J. R., Skofronick, J. G., and Grant, R. M. *Physics of the Body*. Madison, WI: Medical Physics Publishing, 1992.

6-6 Chapman, R. F. *The Insects*. New York, NY: American Elsevier Publishing Co., 1969.

6-7 Cooper, John M., and Glassow, Ruth B. *Kinesiology, 3rd* ed. St. Louis, MO: The C. V. Mosby Co., 1972.

6-8 Cromer, A. H. *Physics for the Life Sciences*. New York, NY: McGraw-Hill Book Co., 1974.

6-9 Frankel, Victor H., and Burstein, Albert H. *Orthopaedic Biomechanics*. Philadelphia, PA: Lea and Febiger, 1970.

6-10 French, A. P. *Newtonian Mechanics*. New York, NY: W. W. Norton & Co., Inc., 1971.

6-11 Frost, H. M. *An Introduction to Biomechanics*. Springfield, IL: Charles C Thomas, Publisher, 1967.

6-12 Gray, James. *How Animals Move*. Cambridge, UK: University Press, 1953.

6-13 Hobbie, R. K. *Intermediate Physics for Medicine and Biology*. New York, NY: Springer, 1997.

6-14 Ingber, D. E. "The Architecture of Life," *Scientific American* (January 1998), 47.

6-15 Jensen, Clayne R., and Schultz, Gordon W. *Applied Kinesiology*. New York, NY: McGraw-Hill Book Co., 1970.

6-16 Kenedi, R. M., ed. *Symposium on Biomechanics and Related Bioengineering Topics*. New York, NY: Pergamon Press, 1965.

6-17 Lauk, M., Chow, C. C., Pavlik, A. E., and Collins, J. J. "Human Balance out of Equilibrium: Nonequilibrium Statistical Mechanics in Posture Control," *The American Physical Society*, 80 (January 1998), 413.

6-18 Latchaw, Marjorie, and Egstrom, Glen. *Human Movement*. Englewood Cliffs, NJ: Prentice-Hall, 1969.

6-19 McCormick, Ernest J. *Human Factors Engineering*. New York, NY: McGraw-Hill Book Co., 1970.

6-20 Mathews, Donald K., and Fox, Edward L. *The Physiological Basis of Physical Education and Athletics*. Philadelphia, PA: W. B. Saunders and Co., 1971.

6-21 Morgan, Joseph. *Introduction to University Physics*, Vol. 1, 2nd ed. Boston, MA: Allyn and Bacon, 1969.

6-22 Offenbacher, Elmer L. "Physics and the Vertical Jump," *American Journal of Physics*, 38 (July 1970), 829–836.

6-23 Richardson, I. W., and Neergaard, E. B. *Physics for Biology and Medicine*. New York, NY: John Wiley & Sons, 1972.

6-24 Rome, L. C. "Testing a Muscle's Design," *American Scientist*, 85 (July–August 1997), 356.

6-25 Strait, L. A., Inman, V. T., and Ralston, H. J. "Sample Illustrations of Physical Principles Selected from Physiology and Medicine," *American Journal of Physics*, 15 (1947), 375.

6-26 Sutton, Richard M. "Two Notes on the Physics of Walking," *American Journal of Physics*, 23 (1955), 490.

6-27 Wells, Katherine F. *Kinesiology: The Scientific Basis of Human Motion*. Philadelphia, PA: W. B. Saunders and Co., 1971.

6-28 Williams, M., and Lissner, H. R. *Biomechanics of Human Motion*. Philadelphia, PA: W. B. Saunders Co., 1962.

6-29 Wolff, H. S. *Biomedical Engineering*. New York, NY: McGraw-Hill Book Co., 1970.

D.2 Chapter 7

7-1 Alexander, R. McNeill. *Animal Mechanics*. London: Sidgwick and Jackson, 1968.

7-2 Chapman, R. F. *The Insects*. New York, NY: American Elsevier Publishing Co., 1969.

7-3 Foth, H. D., and Turk, L. M. *Fundamentals of Soil Science*. New York, NY: John Wiley & Sons, 1972.

7-4 Gamow, G., and Ycas, M. *Mr. Tomkins Inside Himself*. New York, NY: The Viking Press, 1967.

7-5 Hobbie, R. K. *Intermediate Physics for Medicine and Biology*. New York, NY: Springer, 1997.

7-6 Morgan, J. *Introduction to University Physics*, 2nd ed. Boston, MA: Allyn and Bacon, 1969.

7-7 Murray, J. M., and Weber, A. "The Cooperative Action of Muscle Proteins," *Scientific American* (February 1974), 59.

7-8 Rome, L. C. "Testing a Muscle's Design," *American Scientist*, 85 (July–August 1997), 356.

D.3 Chapter 8

8-1 Ackerman, E. *Biophysical Sciences*. Englewood Cliffs, NJ: Prentice-Hall, 1962.

8-2 Hademenos, G. J. "The Biophysics of Stroke," *American Scientist*, 85 (May–June 1997), 226.

8-3 Morgan, J. *Introduction to University Physics*, 2nd ed. Boston, MA: Allyn and Bacon, 1969.

8-4 Myers, G. H., and Parsonnet, V. *Engineering in the Heart and Blood Vessels*. New York, NY: John Wiley & Sons, 1969.

8-5 Richardson, I. W., and Neergaard, E. B. *Physics for Biology and Medicine*. New York, NY: John Wiley & Sons, 1972.

8-6 Ruch, T. C., and Patton, H. D., eds. *Physiology and Biophysics*. Philadelphia, PA: W. B. Saunders Co., 1965.

8-7 Strait, L. A., Inman, V. T., and Ralston, H. J. "Sample Illustrations of Physical Principles Selected from Physiology and Medicine," *American Journal of Physics*, 15 (1947), 375.

D.4 Chapters 9 to 11

11-1 Ackerman, E. *Biophysical Science*, Englewood Cliffs, NJ: Prentice-Hall, 1962.

11-2 Angrist, S. W. "Perpetual Motion Machines," *Scientific American* (January 1968), 114.

11-3 Atkins, P. W. *The 2nd Law*. New York, NY: W. H. Freeman and Co., 1994.

11-4 Brown, J. H. U., and Gann, D. S., eds. *Engineering Principles in Physiology*, Vols. 1 and 2. New York, NY: Academic Press, 1973.

11-5 Casey, E. J. *Biophysics*, New York, NY: Reinhold Publishing Corp., 1962.

11-6 Loewenstein, W. R. *The Touchstone of Life: Molecular Information, Cell Communication, and the Foundations of Life*. New York, NY: Oxford University Press, 1999.

11-7 Morgan, J. *Introduction to University Physics*, 2nd ed. Boston, MA: Allyn and Bacon, 1969.

11-8 Morowitz, H. J. *Energy Flow in Biology*. New York, NY: Academic Press, 1968.

11-9 Peters, R. H. *The Ecological Implications of Body Size*. Cambridge University Press, 1983.

11-10 Rose, A. H., ed. *Thermobiology*. London: Academic Press, 1967.

11-11 Ruch, T. C., and Patton, H. D., eds. *Physiology and Biophysics*. Philadelphia, PA: W. B. Saunders Co., 1965.

11-12 Stacy, R. W., Williams, D. T., Worden, R. E., and McMorris, R. W. *Biological and Medical Physics*. New York, NY: McGraw-Hill Book Co., 1955.

D.5 Chapter 12

12-1 Alexander, R. McNeil *Animal Mechanics*. Seattle, WA: University of Washington Press, 1968.

12-2 Brown, J. H. U., and Gann, D. S., eds. *Engineering Principles in Physiology*, Vols. 1 and 2. New York, NY: Academic Press, 1973.

12-3 Burns, D. M., and MacDonald, S. G. G. *Physics for Biology and Pre-Medical Students*. Reading, MA: Addison-Wesley Publishing Co., 1970.

12-4 Casey, E. J. *Biophysics*. New York, NY: Reinhold Publishing Corp., 1962.

12-5 Cromwell, L., Weibell, F. J., Pfeiffer, E. A., and Usselman, L. B. *Biomedical Instrumentation and Measurements*. Englewood Cliffs, NJ: Prentice-Hall, 1973.

12-6 Marshall, J. S., Pounder, E. R., and Stewart, R. W. *Physics*, 2nd ed. New York, NY: St. Martin's Press, 1967.

12-7 Morgan, J. *Introduction to University Physics*, 2d ed. Boston, MA: Allyn and Bacon, 1969.

12-8 Richardson, I. W., and Neergaard, E. B. *Physics for Biology and Medicine*. New York, NY: John Wiley & Sons, 1972.

12-9 Stacy, R. W., Williams, D. T., Worden, R. E., and McMorris, R. W. *Biological and Medical Physics*. New York, NY: McGraw-Hill Book Co., 1955.

D.6 Chapter 13

13-1 Ackerman, E. *Biophysical Science*. Englewood Cliffs, NJ: Prentice-Hall, Inc., 1962.

13-2 Bassett, C. A. L. "Electrical Effects in Bone," *Scientific American* (October 1965), 18.

13-3 Bullock, T. H. "Seeing the World through a New Sense: Electroreception in Fish," *American Scientist* 61 (May–June 1973), 316.

13-4 Delchar, T. A. *Physics in Medical Diagnosis*. New York, NY: Chapman and Hall, 1997.

13-5 Hobbie, R. K. "Nerve Conduction in the Pre-Medical Physics Course," *American Journal of Physics*, 41 (October 1973), 1176.

13-6 Hobbie, R. K. *Intermediate Physics for Medicine and Biology*. New York, NY: Springer, 1997.

13-7 Katz, B. "How Cells Communicate," *Scientific American* (September 1961), 208.

13-8 Katz, B. *Nerve Muscle and Synapse*. New York, NY: McGraw-Hill, Inc., 1966.

13-9 Miller, W. H., Ratcliff, F., and Hartline, H. K. "How Cells Receive Stimuli," *Scientific American* (September 1961), 223.

13-10 Scott, B. I. H. "Electricity in Plants," *Scientific American* (October 1962), 107.

D.7 Chapter 14

14-1 Ackerman, E. *Biophysical Science*. Englewood Cliffs, NJ: Prentice-Hall, Inc., 1962.

14-2 Blesser, W. B. *A Systems Approach to Biomedicine*. New York, NY: McGraw-Hill Book Co., 1969.

14-3 Cromwell, L., Weibell, F. J., Pfeiffer, E. A., and Usselman, L. B. *Biomedical Instrumentation and Measurements*. Englewood Cliffs, NJ: Prentice-Hall, Inc., 1973.

14-4 Davidovits, P. *Communication*. New York, NY: Holt, Rinehart and Winston, 1972.

14-5 Scher, A. M. "The Electrocardiogram," *Scientific American* (November 1961), 132.

D.8 Chapter 15

15-1 Ackerman, E. *Biophysical Science*. Englewood Cliffs, NJ: Prentice-Hall, Inc., 1962.

15-2 Davidovits, P., and Egger, M. D. "Microscopic Observation of Endothelial Cells in the Cornea of an Intact Eye," *Nature* 244 (1973), 366.

15-3 Katzir, A. "Optical Fibers in Medicine," *Scientific American* (May 1989) 260, 120.

15-4 Marshall, J. S., Pounder, E. R., and Stewart, R. W. *Physics*, 2nd ed. New York, NY: St. Martin's Press, 1967.

15-5 Muntz, W. R. A. "Vision in Frogs," *Scientific American* (March 1964), 110.

15-6 Ruch, T. C., and Patton, H. D. *Physiology and Biophysics*. Philadelphia, PA: W. B. Saunders and Co., 1965.

15-7 Wald, George. "Eye and the Camera," *Scientific American* (August 1950), 32.

D.9 Chapter 16

16-1 Ackerman, E. *Biophysical Sciences*. Englewood Cliffs, NJ: Prentice-Hall, Inc., 1962.

16-2 Burns, D. M., and MacDonald, S. G. G. *Physics for Biology and Pre-Medical Students*. Reading, MA: Addison-Wesley Publishing Co., 1970.

16-3 Delchar, T. A. *Physics in Medical Diagnosis*. New York, NY: Chapman and Hall, 1997.

16-4 Dowsett, D. J., Kenny, P. A., and Johnston, R. E. *The Physics of Diagnostic Imaging*. New York, NY: Chapman and Hall Medical, 1998.

16-5 Hobbie, R. K. *Intermediate Physics for Medicine and Biology*. New York, NY: Springer, 1997.

16-6 Pykett, I. L. "NMR Imaging in Medicine," *Scientific American* (May 1982), 78.

16-7 Pizer, V. "Preserving Food with Atomic Energy," United States Atomic Energy Commission Division of Technical Information, 1970.

16-8 Schrödinger, E. *"What Is Life?" and Other Scientific Essays*. Garden City, NY: Anchor Books, Doubleday and Co., 1956.

Appendix E

Answers to Numerical Exercises

E.1 Chapter 1

1-1(b). $F = 254$ N (57.8 lb)
1-3. $\theta = 72.6^\circ$
1-4. Maximum weight $= 335$ N (75 lb)
1-5(a). $F_m = 2253$ N (508 lb), $F_r = 2386$ N (536 lb)
1-6. $F_m = 720$ N, $F_r = 590$ N
1-7(a). $F_m = 2160$ N, $F_r = 1900$ N
1-8. $\Delta F_m = 103$ N, $\Delta F_r = 84$ N
1-10. $\Delta x = 19.3$ cm, $v = 4$ cm/sec
1-11. $F_m = 0.47$ W, $F_r = 1.28$ W
1-12(a). $F_m = 2000$ N, $F_r = 2200$ N; (b). $F_m = 3220$ N, $F_r = 3490$ N
1-13. $F_A = 2.5$ W, $F_T = 3.5$ W

E.2 Chapter 2

2-1(a). Distance $= 354$ m; (b). Independ of mass
2-2(a). $\mu = 0.067$
2-3(a). $\mu = 0.195$; (b). with $\mu = 0.195, \theta = 28^\circ$, with $\mu = 0.003, \theta = 43^\circ$

E.3 Chapter 3

3-1. $P = 4120$ watt
3-2. $H' = 126$ cm

3-3. $F_r = 1.16$ W, $\theta = 65.8°$
3-4. $T = 0.534$ sec
3-5(a). $R = 13.5$ m; (b). H = 3.39 m; (c). 4.08 sec
3-6. $v = 8.6$ m/sec
3-7. $r = 1.13$ m
3-8. $v = 8.3$ m/sec
3-9. Energy expended/sec = 1350 J/sec
3-10. $P = 371$ watt

E.4 Chapter 4

4-2. $F = 10.1$ N
4-3. $\omega = 1.25$ rad/sec
4-4. $v_{max} = 62.8$ m/sec
4-5(b). $\omega = 3.55$ rad/sec = 33.9 rpm
4-6. Speed = 1.13 m/sec = 4.07 km/hr = 2.53 mph
4-7. $T = 1.6$ sec
4-8. $E = 1.64$ mv^2

E.5 Chapter 5

5-1. $v = 2.39$ m/sec (5.3 mph)
5-2. $v = 8$ m/sec; with 1 cm^2 area $v = 2$ m/sec
5-3. $h = 5.1$ m
5-4. $t = 3 \times 10^{-2}$ sec
5-5. $v = 17$ m/sec (37 mph)
5-6. Force/cm^2 = 4.6×10^6 dyn/cm^2, yes
5-7. $v = 0.7$ m/sec, no

E.6 Chapter 6

6-1. $F = 2$ W
6-2. $l = 0.052$ mm
6-3. $h = 18.4$ cm
6-4. $l = 10.3$ cm

E.7 Chapter 7

7-2. $P = 7.8$ W
7-6. $p = 15$ atm
7-7. $x = 3.8\%$
7-8. $p = 1.46 \times 10^5$ dyn/cm^2
7-11. $a = 1.09 \times 10^3$ cm/sec^2

E.8 Chapter 8

8-1. $\Delta P = 3.19 \times 10^{-2}$ torr
8-2. $\Delta P = 4.8$ torr
8-3. $h = 129$ cm
8-4(a). $p = 61$ torr; (b). $p = 200$ torr
8-5(b). $R_1/R_2 = 0.56$
8-6. $v = 26.5$ cm/sec
8-7. $N = 7.5 \times 10^4$
8-8. $\Delta p = 79$ torr
8-9. $P = 10.1$ W
8-10(a). $P = 0.23$ W; (b). $P = 4.5$ W

E.9 Chapter 9

9-2. $V = 238$ l
9-3(a). $t = 10^{-2}$ sec; (b). $t = 10^{-5}$ sec
9-5. $N = 1.08 \times 10^{20}$ molecules/sec
9-6. No. breaths/min. = 10.4
9-7(a). Rate = 1.71 liter/h-cm^2; (b). diameter = 0.5 cm

E.10 Chapter 11

11-2. $t = 373$ hours
11-3. $v = 4.05$ m^3
11-4. $t = 105$ days
11-5. Weight loss = 0.892 kg
11-6. $H = 18.7$ Cal/h
11-8(b). Change = 22%; (c). $K_r = 6.0$ Cal/m^2-h-C^0
11-9. Heat removed = 8.07 Cal/h
11-10. Heat loss = 660 Cal/m^2-h
11-11. $H = 14.4$ Cal/h

E.11 Chapter 12

12-1. $R = 31.6$ km
12-2. 1.75 times
12-3. $p = 2.9 \times 10^{-4}$ dyn/cm^2
12-6. $D = 11.5$ m
12-8. Min. size $= 1.7 \times 10^{-2}$ cm

E.12 Chapter 13

13-1(a). No. of ions $= 1.88 \times 10^{11}$; (b). no. of Na^+ ions $= 7.09 \times 10^{14}$/m; No. of K^+ ions $= 7.09 \times 10^{15}$/m
13-8(a). no of cells in series = 5000; (b). no of cells in parallel $= 2.7 \times 10^9$

E.13 Chapter 14

14-1. $i = 13.3$ amp

E.14 Chapter 15

15-1. Change in position = 0.004 cm
15-3. For cornea 41.9 diopters; for lens, min power = 18.7 diopters, max power = 24.4 diopters
15-4. $1/f = -0.39$ diopters
15-5. $1/f = -70$ diopters
15-6. $p = 1.5$ cm
15-7(a). Resolution $= 2.67 \times 10^{-4}$ rad; (b). resolution $= 5.33 \times 10^{-4}$ rad
15-8. $D = 20$ m
15-9. $H = 3 \times 10^{-4}$ cm

Index